National Association of Home Builders

Computer Estimating for Home Builders

Published by
The NAHB Member Computer Services
and
Home Builder Press

Home Builder Press®
National Association of Home Builders
1201 15th Street, NW
Washington, DC 20005-2800
(800) 223-2665
www.builderbooks.com

Computer Estimating for Home Builders
ISBN 0-86718-440-X

Printed in the United States of America

Library of Congress Cataloging-in-Publication Data

Coleman, Thomas, 1955–
Computer estimating for home builders / Thomas Coleman.
p. cm.
ISBN 0-86718-440-X
1. House construction—Estimates—Data processing. I. Title
TH4815.8.C654 1998
690′.837—DC21 97-50146
CIP

Cover art courtesy of Timberline Software Corporation

For further information please contact:

Home Builder Presss®
National Association of Home Builders
1201 15th Street, NW
Washington, DC 20005-2800
(800) 223-2665
www.builderbooks.com

12/97 ER/McNaughton 1,200

Contents

Acknowledgments

The National Association of Home Builders would like to thank the many people who helped directly and indirectly to create this reference book.

Thanks to John Geffel, Vice President of Marketing at Timberline Software Corporation, who gave generously of his time, experience, and resources to provide a clear, unbiased manuscript; to Jean Carmichael, NAHB Director of Member Computer Services, who helped shape this manuscript from concept to printed page. Thanks also to Debra Carpenter-Beck, Timberline Software, who worked many long hours on this project. Deb provided many helpful comments and suggestions, and her work turned this manuscript into clear, concise copy. For contributing to the content of this manuscript with valuable comments and suggestions from their own experience in the home building industry, thanks to the following people from Timberline Software: Ammon Cookson, Kelly Hyvonen, Vicki Roberge, Pat Smith, Greg Tipton, and Steve Watt. Special thanks in this regard also go to Dale Henningsen of Skyline Business Systems.

The many construction professionals who gave their time and ideas to this book have made an invaluable contribution. In particular, special thanks to the members and volunteers of the NAHB: the Information Services Committee (ISC) and the Business Management Committee (BMC). Thanks to Bill Post, Chairman of the ISC, and Emma Shinn, Chairperson of the BMC. Also, thanks to Leon Rogers, who was always available with thorough and thoughtful feedback and advice.

Special thanks go to the following people who reviewed the outline or manuscript, or provided verbal comments, for their excellent suggestions on ways to improve this publication: Catherine Koury Aldridge, Building Profits, Raleigh, NC; William A. Allen, consultant, Redmond, WA; Debra L. Baker, Enterprise Computer Systems, Greenville, SC; Tom Benedict, Wayne Homes, North Canton, OH; Keith Brown, Buildnet, Research Triangle Park, NC; Allan Freedman, NAHB; John Geoffroy, Construction Data Control, Atlanta, GA; John Jones, Softplan Systems, Waterloo, Ontario; Stuart M. Jones, Construction and Management, Raleigh, NC; John W. Neidert, Wayne Homes, North Canton, OH; Dottie Piazza, Piazza Realty, Burlington, WA; Gage Pritchard, Gage Homes, Dallas, TX; Joedy Sharpe, Sharpe Homes, Danville, KY; Norman Speaks, Norm Speaks Showcase Homes, Centerville, OH; Jim Sobeck, Enterprise Computer Systems, Greenville, SC; Alan Trellis, RCM Corporation, Columbia, MD; Bob Whitten, Home Builder Advisory Services/Breland Homes, Madison, AL.

Thanks also to the builders Mary Evelyn Hammond, M/I Homes; Eric Schmitt, Schmitt Building Contractors; and Jay True,

Jim Murphy & Associates, who provided information for real-life case studies. To each of them, a special note of appreciation.

This book was produced under the general direction of Kent Colton, NAHB Executive Vice President and CEO, in association with NAHB staff members James E. Johnson, Jr., Staff Vice President, Information Services Division; Adrienne Ash, Assistant Staff Vice President, Publishing Services; Rosanne O'Connor, Director of Publications; Sharon Lamberton, Assistant Director of Publications; John Tuttle, Publications Editor; Kurt Lindblom, Publications Editor; David Rhodes, Art Director; and Carolyn Kamara, Editorial Assistant.

About the Author

Thomas Coleman, a former home builder, is director of estimating for Timberline Software Corporation. In his current position he regularly works with residential contractors to develop and implement estimating software for their businesses. Timberline Software is a leading developer of estimating, project accounting, and property management software for home builders.

Foreword

Estimating is the lifeblood of the construction industry. Custom home builders obtain much of their work through competitive bids. Production builders must complete an accurate bid in order to determine the potential cost of a project and to establish the sales price of a home.

With the advent of personal computers in the early 1980s, builders began converting from manual to computerized accounting systems. Today most builders' accounting systems are on a computer. Typically, the next business function builders want to computerize is their estimating operations. Many of the large builders have already made this move, and small-volume builders are joining the trend. In fact, according to NAHB market analyses, the hottest-selling software in construction is computerized estimating programs.

If you have decided to computerize your estimating system, you basically have two choices.

- Use a spreadsheet like Excel, Lotus, or QuattroPro for your estimating.
- Purchase a dedicated estimating program from a construction software vendor.

A spreadsheet program is usually the least expensive option and offers many applications besides estimating. However, creating estimating templates with a spreadsheet program can require a lot of time. Normally you will have to spend many hours developing forms and formulas before you are ready to complete your first estimate. This process is a valuable learning experience, but builders seem to be constantly tweaking their system to get better performance, and they eventually find that their customized system looks increasingly like a dedicated estimating program.

A dedicated estimating program specifically designed for builders offers a lot of power "right out of the box." Besides convenience, builders usually have two major reasons for choosing computer estimating programs.

- They want to improve the accuracy of their estimating. Computers can store and track vast amounts of data, including prices, forms, and materials, and rarely make addition or subtraction mistakes.
- They want to increase productivity. A good computer estimating program can easily cut estimating time in half or double the amount of work a builder can bid on at any given time.

Several well-established estimating programs are available for builders, so you have a good chance of finding a program that meets your needs. However, before you go looking for an estimating program, there are a few things you ought to know.

Computer Estimating for Home Builders is a good place to start. Written for the builder or estimator who is shopping for an estimating program, this book will help you to define your needs, ask the right questions of vendors, and understand the answers that you receive. It outlines those things you should know when making a final decision and gives you good advice on how to implement the software once you have bought it. In short, it helps you get over the frustration that often accompanies the implementation of any new computer system.

This book does not favor specific estimating programs. Instead, it helps you to filter through all of the information available from software vendors to determine what is most important for you to consider for your own company.

Many builders who have converted from manual to computerized estimating indicate they typically double their estimating productivity while increasing the accuracy of their estimates. A good estimating program can often turn something that you do out of necessity into a very powerful management tool.

Good luck and good estimating!

by LEON ROGERS
Professor, Construction Management
Brigham Young University

Introduction

Entrusting any part of your business to a computer can be stressful. Computerization not only requires extra cost, but also forces you to change the way you do business. *Computer Estimating for Home Builders* was written to reduce this stress. It walks you through all the steps necessary to computerize your construction estimates, including selecting and implementing an estimating system, takeoff and analytical tools, preparing reports, and more. It gives you the information you need so you can move to the next stage: producing breakdowns of anticipated construction costs more easily and accurately

One note about this book's use of the word "estimate." Builders tend to refer to this process of cost breakdown in different ways. While some may call it an estimate, others may think of it as a bid or a budget. For the purposes of this book, we will refer to this process as an estimate. Please keep this in mind as you read through the following chapters.

Chapter 1 offers advice on selecting a computer estimating program. It introduces you to the basic elements common to many computer estimating programs. It then leads you through the process of choosing and buying the right estimating software package, including performing a needs analysis, conducting a product review, choosing a software company, and making a decision.

Once you have decided on a computer estimating program, Chapter 2 guides you through the implementation process. It discusses how to best introduce the software into your company and walks you through the process of developing an implementation plan. This chapter was based on input from builders who have survived successful implementation and will help you avoid some common mistakes.

Chapter 3 discusses advanced computer estimating tools that may further enhance your company's productivity. These advanced tools include digitizer software, assemblies, sequencing capability, and budget management. The chapter closes with a discussion of future trends.

Chapter 4 introduces you to a major advantage of computer estimating: the ability to quickly and easily prepare a variety of useful reports and documents. The bulk of the chapter consists of sample forms that will give you an idea of the variety of computer estimating capabilities.

Chapter 5 discusses the possibilities of linking your computer estimating software with other computer programs. It introduces the concept of integration and covers some of the procedures you eventually may want to integrate: job accounting, purchasing, scheduling, pricing, and computer-aided design (CAD).

Chapter 6 focuses on the practical end of computer estimating. It profiles three home builders who are currently using computer estimating and illustrates the impact of the software on their business.

The book concludes with a glossary of terms, an estimating software buyer's checklist, and a directory of estimating software packages.

Chapter One

Selecting a Computer Estimating System

Basic Elements

When you begin your search for a computer estimating system, you'll notice some general similarities among systems. Most estimating systems designed for residential construction have two main components: a form or spreadsheet look to the software and an easy-to-access database. (See Figure 1.1.) The spreadsheet or form section is where you build your estimate. The database is where you store material prices, trade contractor quotes, and anything else you'll use again and again in your estimates.

An estimating system is typically driven by the takeoff process. A takeoff may involve merely selecting trade contractor names from your database and entering their quotes. However, if you do your own framing or other work, you may do a more complete takeoff to determine your own unit costs. You can also use an estimating system to do a more conceptual square-foot estimate when only preliminary information or drawings are available. Some builders even do comparison takeoffs on items such as masonry to negotiate better prices with their trade contractors. Finally, if you are a production builder or do semicustom work, you may take off your standard plans only once, then use your estimating system to keep prices up to date.

No matter how you estimate, understanding the basic components and functions of estimating software will help you determine the best estimating system to fit your needs.

Available Options

This book covers a wide range of capabilities available from today's estimating systems. Some of these features you will need; others you will not. Fortunately, there are many estimating systems to choose from—whether you need an affordable system that covers basic estimating functions, or you need much more advanced capabilities.

To successfully select a computer estimating system, you need to understand your requirements and staff resources, and have a realistic budget. Use the process described in this chapter as a

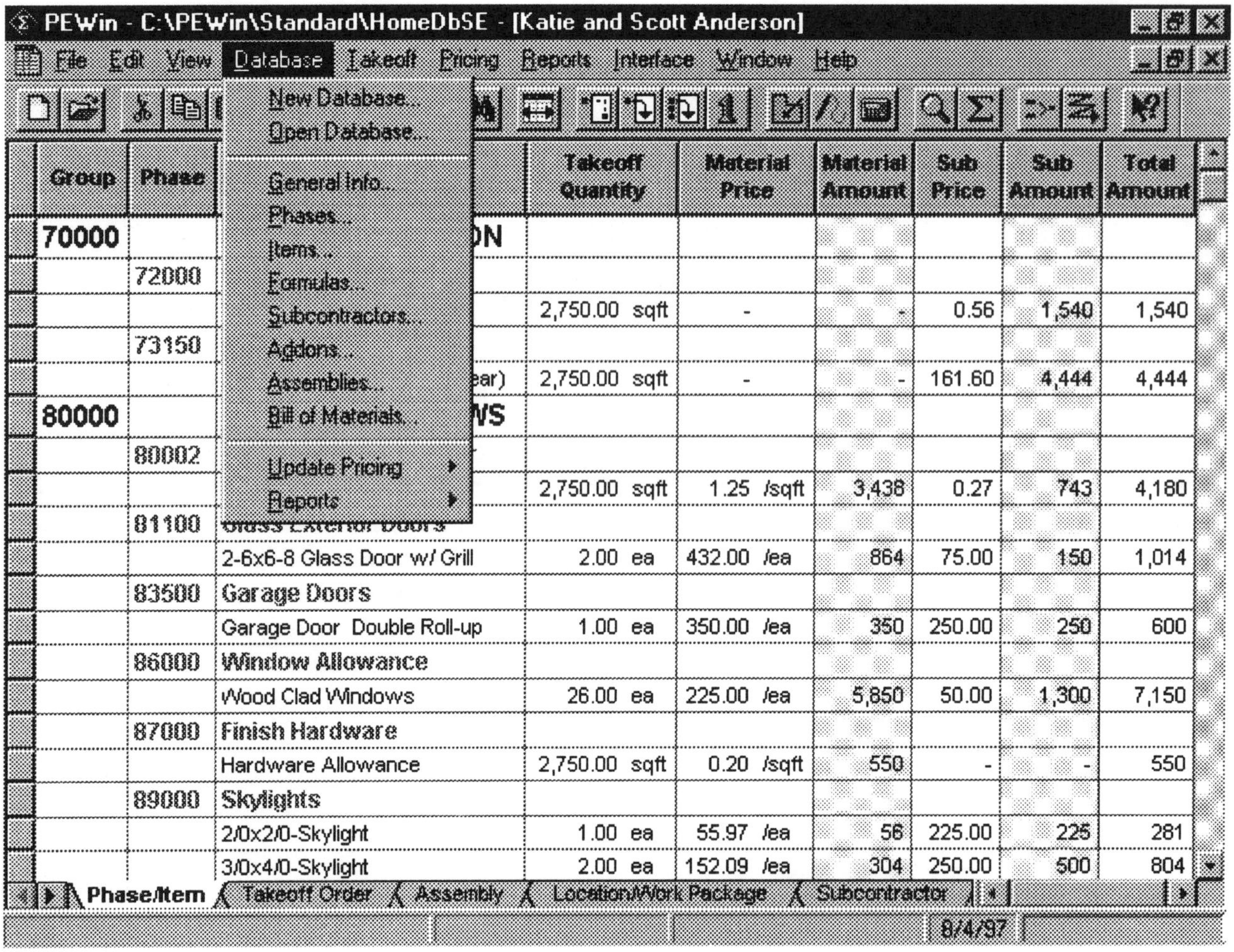

Figure 1.1 Two Components of Estimating Systems: Spreadsheet and Database

guide to select a computer estimating system that is right for your company. (See also Appendix B.)

Needs Analysis

To design and build someone's dream home, you need to know something about their lifestyle, design tastes, and other preferences. The same is true for computer estimating. To choose the system that's right for you, you need to perform a needs analysis.

Even if your company is small, start your analysis by talking with every employee involved with estimates and purchasing. A major reason for these discussions is to get everyone's support for the new estimating system. People's attitudes about change and the inconvenience of new procedures can directly affect the success of a project. By discussing the purpose and goals of the new procedures, you will allow everyone to participate in the decision and feel comfortable, rather than threatened, by the change a new system will inevitably bring.

Your accountant, in particular, should be included early in the selection process and during implementation. Not only is your accountant usually aware of various estimating systems, but he or she can judge which systems will supply necessary reports and fit most easily into your method of operation.

As you perform your needs analysis, include even the most minor need. Record all requirements, and when the list is complete, prioritize each need according to its overall importance. Consider technical as well as personal requirements.

Develop a Needs List

Brainstorm what you think your company needs from a computer estimating system. Consider what estimating functions you want to improve and why. The list in Figure 1.2 will get you started. It outlines some of the needs often considered by home builders looking for a computer estimating system. Some items may not apply to you, or you may identify other needs that are not listed.

When brainstorming your needs, you might want to look through the Estimating Software Buyer's Checklist. (See Appendix B.) It may suggest ways to use computer estimating that might not have occurred to you. Keep in mind that your business is not static. Change is inevitable as your company continues to develop, so look beyond your immediate needs. Think about what you may need from a computer estimating system next year, or even two years from now.

Also, remember you can find computer estimating systems that integrate with other business functions, such as job costing and accounting, purchasing, scheduling, and computer-aided design (CAD). Chapter 5 discusses the benefits of integration in detail and contains information that will help you select the best system for your company.

Prioritize Each Need

Next, group your needs in a logical order and prioritize them. Identify which estimating system features you must have and which you would like to have. Refer to Appendix B for a checklist of system features that may be available from different products.

After prioritizing your needs, you must determine which estimating system comes closest to matching them. If you feel you lack the time and technical expertise to do this, consider hiring a computer estimating consultant. Should you hire a consultant, select someone who understands construction and is familiar with several computer estimating packages, not just one.

As mentioned before, think ahead to what your needs may be in the future. You don't want to purchase a system that your business quickly outgrows.

Figure 1.2 Potential Needs List

The answers to the questions below will help determine the system your company needs. Use the margin and space at the bottom for notes and additions to the list.

_____ What type of homes do you build (e.g., standard models with options, semicustom homes, or highly customized homes)?

_____ Do you obtain work through competitive bidding, or do you set your own sales price?

_____ Do your estimates differ tremendously from one job to the next, or are they fairly similar? How standard are your plans?

_____ Who currently produces your estimates, and what procedures do they follow? Do you want to maintain some or all of those procedures? Which ones would you like to improve upon?

_____ Do estimate procedures and practices shared by multiple personnel need to be standardized in your company?

_____ Do you need to more accurately track the average unit cost of your homes (e.g., cost per square foot of slab, or cost per linear foot of interior wall)?

_____ Do you need to provide quick preliminary estimates to prospective clients? Who delivers these estimates—your sales staff or the person who prepares the estimates? How quickly do you want to create these estimates? How automated do you want this process to be?

_____ How quickly do you want to present final estimates to clients?

_____ How do you apply sales tax, general overhead, and profit to the job? Do you want to embed these and other mark-ups in your estimate or show them separately?

_____ What type of organizational structure do you use for your estimates?

_____ How much detail do you want in your estimates?

_____ Do you need or want to take off your own quantities for lumber or other materials?

_____ Do you want to use a digitizer to enter plan dimensions into the estimating system by tracing over blueprints?

_____ Do you need material-waste factors to be easily adjustable (e.g., you may figure 5 percent waste on lumber, but only 4 percent waste on concrete)?

_____ If you have your own crews, do you want to calculate cost based on crew size and number of hours?

_____ Do you use purchase orders?

_____ Do you want to customize units of measure to match your purchasing and accounting systems (e.g., if you buy lumber in thousand board feet, you will need a system that accommodates four decimal points)?

_____ How do you handle change orders? What would make this process easier? Do you want your change orders to revise your original estimate?

_____ What reports do you need?

_____ Who needs reports, and in what format?

_____ What types of reports do you want to provide to clients, and when?

_____ Do you use cost information for negotiating the sales price with clients?

_____ What type of cost breakdown do you want to provide the bank for construction loans? In what format?

_____ What documents do you need or want to provide to suppliers (e.g., quote sheets)?

_____ Do you want to convert estimates into bills of material? Do you want your estimating system to interface with other software programs? Select which programs:

- _____ Job cost accounting
- _____ CAD
- _____ Purchasing
- _____ Scheduling
- _____ Sales contact management
- _____ Word processing
- _____ Generic electronic spreadsheets
- _____ Presentation software
- _____ Other

_____ How familiar are you with computers? How computer-literate is your staff?

_____ How much and what kind of support will you need? (Be realistic.)

- _____ One-on-one consulting
- _____ Training classes offered by software companies and their local representatives
- _____ Assistance over the phone
- _____ On-line assistance through a web site or software help system
- _____ User group

_____ Do you have data—such as material items, units of measure, trade contractor names—in electronic format that you'd like to transfer into the new estimating system without rekeying?

Decide How Much to Invest

Computer estimating systems can range in price from a few hundred dollars to thousands of dollars. Before you spend weeks researching systems, decide how much you are willing to invest in direct and indirect dollars over a specific period of time.

When you consider this figure, amortize the cost over a period of several years. As you learn more about your needs and about what is available, you may wish to adjust this figure. Start with a realistic range to avoid wasting time researching systems that may not be financially feasible for you at this time.

Keep in mind that hardware and software are only part of the cost of a computer estimating system. To realize the full benefit of the system, you should be prepared for the "soft" costs associated with the time, training, and support it will take to initially learn and implement the software, as well as to maintain it long term.

The good news is that, when implemented correctly, computer estimating can quickly improve your estimating productivity, even within the first few weeks of using the system. (You should expect, however, a slight productivity decline at the very start of implementation). After your system has been up and running for awhile, you can realistically create at least twice as many new estimates as before. This productivity increase should be considered when determining the true cost impact of computerizing your estimating process. To better understand this impact, use your own figures in the following example to get a realistic idea of your cost per estimate before and after computerization:

Example

Cost per estimate—manual method	
Annual cost to prepare estimates	$50,000
Number of estimates produced annually	25
Cost per estimate	$2,000
Cost per estimate—computerized method	
Annual cost to prepare estimates in first year of computerization	$50,000
Amortized computerization costs (amortization terms = 3 yrs)	$10,000
(includes computer hardware support and upgrade fees)	
Twice as many estimates can now be produced	50
Cost per estimate	$1,200

Although it is hard to put a dollar figure on the ability to respond more quickly and accurately to your clients' cost questions, keep this ability in mind as an additional reason for investing in computer estimating. You should also consider that many estimating systems will automatically set up your job cost budget, eliminating yet another time-consuming task.

Learn the Jargon

When researching computer estimating systems, you will probably come across some new terms. Assume nothing when doing your research. Never hesitate to ask in a demonstration, "What does that mean?" At the very least, become familiar with the terms described in this book's glossary. (See Appendix A.) Knowing these terms will help you to better communicate your needs during a demonstration. But be aware that the same term can have different meanings in different computer estimating packages. Conversely, the same feature may have different names in different packages.

Product Review

After you have done your homework and decided what you need, it is time to find out more about the products that are available. Appendix B provides a checklist of features to help you compare products and software companies. The following are suggestions on how to select products for review.

Talk to Users

One of the best ways to identify products to review is to contact builders who are using computer estimating. Talk to both management and those who actually work with the system. Ask them why they decided to use computer estimating, what package they selected, and why they chose that package. Compare their operations with your own and note any similarities. What computer equipment do they use? Who uses the system? How long did it take to see an increase in productivity? What do they estimate their payback period was or will be? Does their computer estimating system integrate with a job cost accounting, purchasing, scheduling, or CAD system?

Be sure to ask what they like and don't like about their computer estimating system. Ask what they would have done differently if they could do it over again. You might also ask about other systems they evaluated and why they did not select those systems.

Keep in mind that software companies upgrade their products frequently. Features that a user describes may be outdated or new features may have been added. Take notes on whom you speak with and what system they are using. After you have seen a demonstration, you may want to talk to some of these people again. If they mention something you overlooked during your own needs analysis, add it to your needs list.

Read the Literature

Collect and read marketing literature and other periodicals on systems offering the features that address your needs. Some software companies advertise in building-trade magazines. To save time, call the company to request product-specific literature, or investigate their web site on the Internet. Some magazines publish product reviews or lists of features that give a general overview of the system. The National Association of Home Builders (NAHB) Home Builder Press can also provide information on computer estimating software through its publications.

Attend Trade Shows

If you plan to attend the national NAHB Convention or any building trade shows scheduled in your area, find out which companies are exhibiting and what seminars on computer estimating they are offering. This is an ideal way to collect a great deal of information quickly. Decide which companies you want to visit before you go, and bring your needs list with you.

NAHB presents hands-on computer training workshops at its annual convention, as well as at regional conferences and special events. The workshops are an excellent opportunity to try out some of the latest products on the market.

Narrow the Choices

After you have selected two to four packages that you feel fit your needs, you are ready for product demonstrations. Call the software company and get the name of the closest certified representative or retail outlet. Some companies may also sell to you direct, either by phone, mail order, or the Internet.

Arrange for a demonstration by calling the local rep or whomever the company recommends. During this call, briefly discuss your operation, what some of your needs are, and what you expect to see. This will help gear the demonstration to your specific requirements. Send the appropriate people to the demonstration and plan to spend two to four hours there. Bring your needs list and one of your typical estimates.

Ask for Product Demonstrations

Builders who use computer estimating systems can sometimes give you the best demonstrations. But even if you know the builder well, this is a lot to ask, as a meaningful demonstration will take several hours of the builder's time. First ask a software company's local representative to come to your place of business for an introduction to the system. Such an onsite visit allows the software rep to view and address your specific business needs.

Before the demonstration starts, describe for the software rep your level of computer estimating experience, give an overview of your operation, and list your basic requirements. This will help the rep gear the demonstration to your company's needs. Let the software rep demonstrate the full system, but don't hesitate to ask questions at any point. Ask a lot of questions, but avoid getting sidetracked on specific details too early in a demonstration. If you do, you may not get a clear overview of the system's capabilities. Instead, you may focus too closely on one or two features it does or doesn't do well.

After the presentation, pull out your list of requirements. Discuss each one with the rep and take notes to refresh your memory later. Don't expect to learn everything you need to know watching a demonstration. If possible, try doing a few operations yourself. If anyone in your group has computer estimating experience, you might ask him or her to spend some time experimenting with the system. Inquire about trial software, which is offered by some companies.

Some computer estimating systems have optional programs or capabilities that can be purchased to perform specific tasks. Ask about the availability of these programs. More important, ask if the system shown has any additional software loaded and used in the demonstration. You should know exactly what you are looking at. If the answer is yes, ask what benefits or additional capabilities these optional programs provide and whether you should consider them for your needs.

If you are interested in integrating your estimating system to other applications, such as job cost accounting and purchasing, be sure to read Chapter 5 of this book. During the demonstration, ask to see how such an interface would work. Although the additional capability may exist, you should analyze how easy it is to perform the interface function and whether the benefits are worth it.

If a specific operation is important to you, ask not only if it can be done but also how easy it is to accomplish. Then have the software rep demonstrate it for you. Beware of the salesperson who quickly answers yes to all your questions. Seldom does one system have everything you would like. Most likely you will have to decide which features are the most important to you and trade off against the limitations.

Although most software is continually being updated and enhanced, don't base today's buying decision on some feature that is talked about in future tense. Until you can actually see how something works, you'll never know if it can do what you need to have done.

During the demonstration, discuss the hardware required both to run the system and to output your estimates, reports, and other documents. Although hardware options generally can be upgraded or added later, it is often more cost-effective to start with the best hardware setup.

Ask if the product has different levels. Some product lines offer a basic system with optional, more advanced features. Also determine if the system comes with a database. What does that database include? Can it be easily modified to fit your specific needs? With some estimating systems, you can also purchase larger databases from the software company itself or other database developers with experience in residential construction.

Before the software rep leaves, ask for the names of home builders in your local area who are using the system. Call them to discuss specific questions that you may have.

Choosing a Software Company

To find the best product for your operation, you should investigate the software company as best you can. When considering a company, keep these factors in mind.

- Length of time in business.
- Financial stability. If the company's stock is publicly traded, request the most recent annual and quarterly financial reports, as well as a history of the company's financial performance. Financial information services can also provide financial histories for nonpublic companies.
- Number of customers in residential construction or number of current installations.
- Number and background of support staff. You may even want to call the support staff before buying the software to see how they can answer specific questions and how long it takes.
- What is included in technical support, training, and documentation (manuals, on-line help, tutorials).
- Release date of the current product version.
- How frequently the program is updated and how updates are handled. Ask for a history of past releases.
- Company's plans for the software's future.
- If applicable, who represents the company locally.

Training and Support

Training and support are important factors that will have a direct impact on your success with a new system. Do not plan to learn everything about a system from the manual and on-line help system or you may waste weeks of valuable time.

Inquire about training classes. Contact the software company and your local rep, trade schools, or other software-training companies in your area. Find out about classes for beginning and advanced users. You may even want to attend a class *before* buying the software to gain a more thorough understanding of the system's capabilities. Ideally, you want a training program that has been certified by the software company. Training should deal specifically with your estimating application as well as with generic estimating techniques. Get details on the cost of training and when courses are offered.

Inquire about both telephone and onsite support and the costs of each. Does the software company or its local rep offer support contracts or are you charged per call or visit? Is a toll-free support number available for you to call when you have questions? Ask how many full-time support people the company has on staff, and about their backgrounds. Find out if the company offers on-line support (for example, on the Internet or fax support services). Are there any other costs that will be incurred? What do other users of the software say about the quality of the support and training they've received?

Ask to see the product documentation or on-line help system before you make a purchase, and take it for a test drive. Try to look something up and then follow the instructions. Was it easy to find what you were looking for? Did you understand the descriptions and steps provided? Some software companies also provide helpful introductory booklets or tutorials that give you step-by-step instructions on how to get up and running from square one.

Making the Decision

After you've reviewed the software company, its products and services, and the hardware required, move forward with your decision while the information is still fresh in your mind. By waiting too long to make a decision (even if you decide not to buy at this time), you risk forgetting or overlooking important facts, or having the software change. Make sure all those who will use or be affected by the software have had a chance to review it from their perspectives. As a final step, you may want to contact a user again with more specific questions.

If you plan to purchase the hardware and software from the same source, you should request a proposal that includes the following:

- Cost for low-, medium-, and high-end hardware alternatives, with descriptions of what the actual performance differences will mean to you (for example, speed and better screen image)
- All software costs broken down by each component (if any), with a description of the functions each performs
- The cost per station or estimator
- A copy of the software-license agreement
- The cost for setting up and installing the system, and details on what setup covers
- The amount of initial training provided and the cost of this training (if additional classes are offered, find out when and where they are held, and the cost)

- Details on the hardware warranty
- The cost of maintenance for all proposed hardware, and the details of any maintenance contract offered
- The cost and specific details for telephone or on-line support
- The company's policies regarding upgrades when a new software version is released
- The cost of an annual maintenance fee for the software (if one exists)
- An installation timetable that details responsibilities (the software company's, its local representative's, and yours) from the time the order is signed

If you are purchasing hardware separately from software, you should still get answers to the questions described above—although the answers will come from different sources. Be sure to find out who is responsible for installing your computer estimating software and ensuring that it works with the hardware.

New technology, such as computer estimating, can provide your business with valuable tools to gain efficiency and productivity. The key to their success in your organization is to research your needs carefully, select the system that best matches your needs, and then commit yourself to its proper implementation.

Chapter Two

Implementing a Computer Estimating System

After you've selected a computer estimating system, you must orchestrate its implementation within your company. That task requires careful planning and patience. This chapter offers a blueprint for implementation. The material is based on input from builders who have gone through successful implementations.

While some of the information discussed is more appropriate for medium to large home builders, many of the strategies and techniques can be applied to home-building companies of all sizes.

Plan Ahead

Computerizing your estimating process can be one of the most rewarding steps you take for your company. But the introduction of any new technology may also require changes in your business.

Your operation has evolved over time to provide for your company's needs. You have developed procedures that enable a smooth workflow. Computer estimating will provide you with an opportunity to reevaluate your estimating procedures. Planning for changes in procedures before introducing computers can protect your company from costly mistakes in selecting a system and can significantly reduce confusion that may occur during the transition.

Keep in mind that implementation is not an isolated process that rests solely with those who prepare your estimates. You need to involve those who handle your accounting, as well as field staff and anyone else whose work will be affected by the new estimating system. These people should participate in planning procedural changes and must be aware of and support the timelines and benchmarks you establish for implementation.

To successfully implement a computer estimating system, be sure you have the following:

- A computer estimating system manager and implementation team (even if that consists only of you, as the owner, and your bookkeeper)
- An implementation plan
- Involvement from management, accounting, and field staff, as well as those who will directly use the system
- Commitment to the computer estimating system from all involved (this includes allotting the time needed to complete the implementation)
- Flexibility

Of course, some companies use their computer estimating system productively without meeting some of these conditions. For example, they may have started without a formal plan. But whether written or not, a plan evolved. Or they may not have created the position of computer estimating system manager, but someone became the leader by virtue of ability. However, taking the time to select the manager, involve others, and prepare the plan will greatly reduce the unproductive time associated with implementing your new system.

The System Manager

If you are a larger home builder, selecting a computer estimating system manager is the first step to take after deciding to computerize your estimating operations. He or she will not only help in choosing the right system, but will greatly influence the transition to computer estimating. The estimating system manager is responsible for writing a computer estimating plan, implementing it, and changing the plan as experience dictates. In short, he or she will be the champion of computer estimating in your company. Because this is a complex task, certain qualifications for the position are essential.

While implementation may not be as complex if you are a smaller builder, you still should make sure that the main responsibilities of the system manager are covered—either by you or by another person, such as a consultant.

Management Skills

Because the computer estimating system manager may supervise the computer-related activities of others, he or she should have some management skills. Employees who have considerable experience in preparing estimates manually or with generic electronic spreadsheets may feel threatened by new computer technology. The introductory period can be a critical time for these people. The system manager must carefully guide them through the implementation process.

Another aspect of the position that requires skill is managing the expectations of managers. Those who depend on the timely output of estimates will try to press for the fastest possible implementation, which may not be in the best interest of long-term computerization goals. The computer estimating system manager must defend the group against attempts to rush progress.

Communication Skills

Good communication skills are required for dealing effectively with both staff and management. Furthermore, the computer estimating system manager will have to create and justify the computer estimating plan for upper management and explain the practical details of implementation to the computer estimating users. It is not enough that the system manager understand computers and computer estimating. He or she must be able to share that understanding with others.

Estimating Experience

Computers in general, and computer estimating in particular, are simply tools. Knowledge of computers is definitely an asset to a computer estimating system manager. But it is an asset that he or she can acquire. The prospective manager must be willing and able to acquire these skills.

Far more important than experience with the tool is experience with the job for which the tool is used. The system manager's ability to incorporate the new software into your estimating workflow will directly influence the degree of productivity you will see when computer estimating is implemented. If he or she does not understand your company's current estimating methods, the project is unlikely to succeed. The depth of understanding necessary for the position is likely to be found in a person who has experience with your company's estimating process.

The Implementation Team

In some of the most successful implementations, the estimating system manager has put together a formal implementation team. This not only takes some of the burden off the system manager, but also provides valuable ideas from a variety of other people in the organization. The team may include members from outside the estimating group, but should include staff who

- Are comfortable with the computer
- Have been or will be trained on the software
- Have adequate estimating experience
- Are allowed time away from other duties to devote to implementation
- Have good communication skills

Consultants

If no one on your staff has the skills and training to select and implement a computer system, hiring a consultant who specializes in this work may help you make a better decision in the long run. A good consultant doesn't add to your overhead costs and can save you hundreds of unproductive hours—and headaches—by providing valuable expertise.

When looking for a consultant, select someone who has knowledge of estimating for residential construction and is abreast of the latest technology in that area. Also important is the consultant's ability to communicate, listen, and effectively manage both his or her time and that of the client. Always ask for and check references to find out how the consultant has performed on projects similar to yours. Ask if the reference would hire the consultant again.

Many estimating software companies maintain a list of certified consultants. When considering these consultants, make sure they have experience implementing systems for home builders on top of expertise with the software. Ask the software company for the qualifications their consultants are required to meet to become certified. You can also locate consultants through professional associations, referrals from other builders, conferences, local community colleges, and professional articles and journals.

Once you've made your selection, the consultant should provide you with a contract that clearly defines project expectations and scope. The contract should outline exactly what you are paying for. Throughout the project, expect the consultant to make recommendations from both a procedural and a software perspective. And when software implementation is complete, expect those recommendations to lead to a more productive staff, improved management of your company, and a more profitable bottom line.

Commitment

It is not enough to be interested in computer estimating. A company must be willing to devote time, energy, and money to make implementation work. Too often, software packages are left to gather dust on a bookshelf because no one took adequate time to learn how to use them properly and productively.

Reaching your final implementation goal won't happen overnight. It will take time to fully establish the items and prices you want in your database, and users of the new system will need time to learn and adapt to the new environment. Management must be aware of this start-up period and be willing to abide by the computer estimating system manager's decisions about the rate at which the new system is implemented. During this critical time, support from management can speed the integration of computers into the company.

A commitment to computer estimating means a willingness to learn new procedures and to work in a new environment. Each employee must be prepared to work with the estimating system manager to investigate the most effective means of using computer estimating. Some employees could be encountering computer estimating for the first time and will have to devote time and energy to learn how to use the new system and how best to apply it to the job.

There may be other sources of stress as well, including maintenance and security of the database. Good management can minimize stress, but commitment to growing with the new technology is required on every level.

The Implementation Plan

Would you ever build a home without a blueprint? As a builder, you know you could easily start building a house today without a plan, but somewhere along the line you'll run into major problems: delays waiting for materials you just realized you needed, a roof that doesn't sit right because of the way you poured the foundation, etc. The same is true for computer estimating. You can easily set up your system and start producing partial or even full estimates. But without a well-defined plan up front, problems and unnecessary delays will occur. In short, creating a plan is a critical step to a successful implementation.

The first step in creating a plan is to determine what you want to achieve with your estimating system and then set a realistic deadline for reaching that implementation goal. If your budgets typically consist of only 20 or 30 trade-contractor quotes, it can take just a few days to implement your system. If you want much more detailed budgets, full implementation could take several months. However, even with a longer implementation, you can still use your system and see early productivity gains if the implementation is handled in stages as described later in this chapter.

Milestones

A good plan also sets aside time for system installation, staff preparation, initial review of the system, training, ongoing database development, and management. (See Figures 2.1 and 2.2.) Establishing standards and including any integration plans with

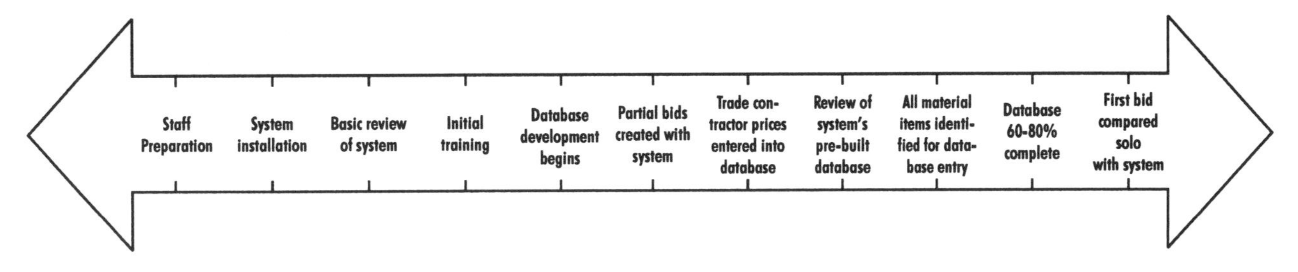

Figure 2.1 Custom Home Building: Computer Estimating Implementation Sample Schedule

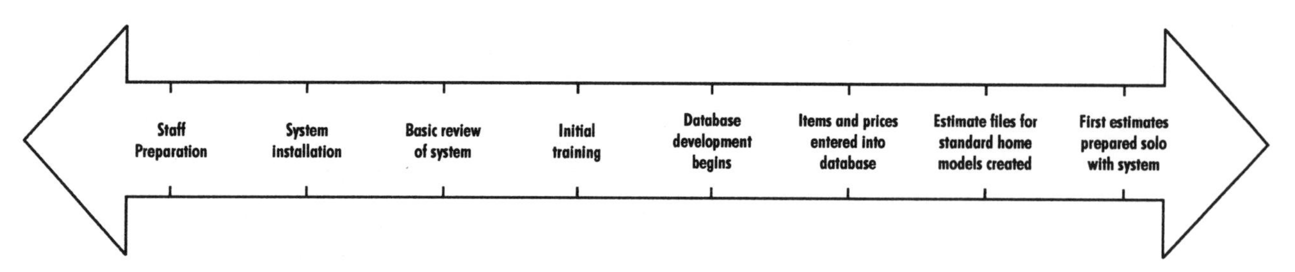

Figure 2.2 Production Home Building: Computer Estimating Implementation Sample Schedule

other business functions are also key steps. Milestones to gauge the progress of your implementation are a critical part of the plan. These milestones can vary depending on the type of homes you build.

Custom Home Building

With custom homes each estimate is unique. Therefore, the implementation goal is to create a database containing a variety of commonly used materials and services that can be quickly compiled into a customized estimate. Some typical implementation milestones for custom homes include the following:

- Partial estimates developed using the system. For example, you may first use your estimating system to prepare your prebuilding checklist of permits (including prices), architectural fees, land cost, and utility hookups. Then you may move on to another section of the estimate. This is a good way to quickly realize time savings with your system, as well as to become familiar with its capabilities without becoming too overwhelmed by the new system.
- Trade contractor lump-sum or unit-price items entered into the database, making them available to prepare estimates.
- All items typically used in homes identified for inclusion in the database.
- Completed review of database items preloaded with your software to determine which items can be used for your own customized database.
- Database 60 to 80 percent complete. At this point, you could begin to prepare full estimates, filling in any gaps as you go. Some systems allow you to enter items to the estimate sheet as you work and then save them to the database if you choose.
- First full estimate completed using only the system. Prior to this, the system is typically used in parallel with existing estimating procedures to ensure everything is working correctly.

Semicustom and Production Home Building

With semicustom and production home building, the estimating system is used to create standard estimates for each home model, including options. Once the estimate is created, the software is used primarily to keep current the construction costs and sale prices for each model. Typical milestones for production home building include the following:

- Vendor prices for items such as windows and doors entered into the database. Trade-contractor information and prices also entered.
- Standard estimate template files of all home models and options completed.
- First estimates created using the system in parallel with your current method.
- First estimate prepared using only the system.

System Installation

Most computer estimating vendors or their local representatives will assist you by ensuring that the software is installed on the computer and by setting up the system in your office. Few vendors will assist in planning the computer estimating system office environment. Proper attention to equipment layout, light sources, furniture, and the work environment can significantly improve operating efficiency. The power supply should be tested and, if necessary, altered to fit the computer's needs. All these factors should be considered in the implementation plan.

Staff Preparation

Staff preparation can begin prior to installation. Every person who prepares estimates should become acquainted with the basic capabilities of the system and the concepts of computer estimating. This can be accomplished by providing them with product literature, articles on computerization, and other material. Staff training on a system prior to installation is not advised, because there will be no system on which to practice.

System Review

Once the software is installed, your staff should spend a week or two to understand its overall workflow. This is the time to practice with the software. Try to create a simple budget and investigate the capabilities of the software's spreadsheet or budget form. You may also want to look at how totals are handled and try to print some reports. During this review process, don't try to understand the mechanics of the software. That will only slow you down. You just want to get a feel for how the software is structured and what it can do.

This review process is an important step to take before anyone goes to training. Even a very basic understanding of the system will provide the background needed to get the most out of a training class.

Training

Builders who have implemented a computer estimating system will tell you that access to training is essential, especially during the initial stages of implementation. Your staff may be able to use the software without training, but you are unlikely to reap the full efficiency and productivity gains of an estimating system unless staff attend training schools and seminars. If you purchase training services from the software company, look for a company with professional training personnel, not just a salesperson doubling as a teacher. User support groups and conferences are also an excellent way to provide ongoing training and information exchange for you and your staff.

Although training is critical, keep it in proper perspective. Your staff will need time and hands-on experience immediately after the training to best apply what they have learned. The longer they wait to implement information from class, the less they will recall.

Establishing and Maintaining the Database

An important step in implementing computer estimating is creating the pricing database. This can take a day or two if your database consists of trade-contractor pricing only. It could take several months if you want your estimates to be much more detailed and enable you to, for example, produce a materials list.

As mentioned in Chapter 1, the database is one of two main components that make up a computer estimating system. Essentially, the database is an electronic price book containing everything you need to build your home. When you want to create an estimate you select the materials, trade contractors, formulas, and other items you need from the database and copy them into the software's spreadsheet or estimate form.

The key to a good database is a coding scheme that helps you find things quickly and easily. The most important aspects of an efficient scheme are

- Meaningful codes
- Consistency

One useful way to structure your database is by the stages of construction. The NAHB Chart of Accounts, which includes codes by construction phases, can be a helpful guide.

If your company already has a coding scheme, this may be a good time to make some improvements. As you devise a new scheme, be sure everyone agrees on the plan before you begin building your new database. Coordinating your database efforts with those involved with job costing, purchasing, and even scheduling is critical to streamlining your company's overall workflow. Be careful, however, not to force job cost codes into the primary structure of your estimating database. Your estimating and job cost codes should be complementary, but not exact duplicates, because they serve two different purposes.

Before you set up your database, you also need to answer some key questions:

- How will you build the database? Do you have to create it from scratch? Can you save time by modifying an industry-standard database? (Some software companies provide ready-built databases that can be easily modified.)
- Do you have an already existing database that merely needs to be imported?
- Who should be responsible for developing the database?
- What is the most efficient way to develop your database?
- How should you handle database security?
- How should you maintain the database?

Because your entire operation depends on the content and accuracy of your data, answering these questions and developing accompanying policies is time well spent. The issues behind these questions are addressed briefly below.

How Will You Develop the Database?

The most time-consuming way to build your database is to create it from scratch. This may be your only option if your type of construction or company budgeting procedures are unique.

Before you begin any database development, however, print out the contents of the database that usually comes preloaded with your software. In some cases it may contain the majority of what you'll need; in other cases, it will have only a small percentage of what you need. More comprehensive residential databases are also available from some software companies or estimating database developers for an additional cost.

These prebuilt databases typically include material items, formulas, and other built-in information that you can easily tailor to your particular needs. In most cases, a prebuilt database can save you a substantial amount of time setting up your computer estimating system. Realize, however, that these databases are based on the individual construction experience of the database developer. Therefore, you may see differences from the way you build homes. Always check the formulas and other information in the database and modify anything that doesn't apply to your own work.

If you already have material items and other estimating data stored in a different software program, you can also save time by transferring the data electronically. To do this, your current software must have export capabilities and your new software must be able to handle the importation. Again, implementation of your new system is a good time to reevaluate the effectiveness of your current estimating information and to make any improvements. Be careful, however, not to tackle too many improvements at once.

Who Should Be Responsible for Developing the Database?

To guarantee a smooth implementation, you should appoint someone to lead the database development. That person may be the estimating system manager, or perhaps he or she will select someone from the implementation team. Some home builders use in-house staff entirely to build the database, drawing on their knowledge of the company's procedures and policies. If you're short on staff time, however, hiring a database development consultant may be a good option. But you will still need to provide information to the consultant so that the database meets your needs.

What is the Most Efficient Way to Develop Your Database?

Many builders find it useful to build the database in stages. They get their system up and running and begin realizing the benefits sooner by focusing initially on the data they use the most. For example, if you do your own lumber takeoff, you may want to focus on including those items in the database first.

Some estimating systems also allow you to build your database as you go. Basically you would add your items directly to the estimate and then save them to the database for future use. By using this approach, your database could conceivably contain about 90 percent of the items you need after just a few estimates.

How Should You Handle Database Security?

To ensure database security, you must address two important issues. First, you must decide who may authorize changes to the database. Too many people making changes can easily result in inconsistent coding, misinformation, and incorrect pricing. To avoid confusion and the potential for error, it is wise to have a single person responsible for modifying the database. Many builders require that all changes be approved by the computer estimating system manager.

Second, you must decide who may have access to the database. Some software packages offer a password feature to allow restricted access. Make sure the password is available only to those who have authorized access.

How Should You Maintain the Database?

Your database will need to be updated periodically. How often you do that depends on regional market and economic conditions. In fact, some home builders may include items but not prices in their database because prices fluctuate too much.

Some software systems provide tools to make database updating more manageable. For example, you may be able to globally adjust prices for all concrete items by a certain percentage. And if you use a pricing service, you may be able to electronically import the most up-to-date prices into your system.

Regardless of how and when you update your database, make sure someone is responsible for this task because it is important to ensuring the success of your estimating system.

Office Procedures

One benefit of computers that is largely overlooked by potential users is that they make you think about current procedures more critically. Although every office has different procedural requirements, different detail standards, and so on, standards can be included in the implementation plan for the following areas:

- Estimating procedures
- Database coding organization
- Estimate files, including file names
- Estimate breakdown structure and organization
- File storage, including system backup, frequency, and retrieval
- Outputs, such as reports

Integration with Other Functions

If you plan to integrate your estimating system with other applications, such as job cost accounting or purchasing, you should include this in your implementation plan. But be realistic. It's best to integrate with other systems once your computer estimating system is fully operational. Trying to change the methods of several job areas at the same time can disrupt workflow. Chapter 5 discusses integration of computer estimating with other functions.

Stay Flexible

If all companies were the same, it would be easy to introduce computer estimating. There would be standard systems and standard books of procedures. But all companies are not the same, and everyone who uses computer estimating has had to learn the process the hard way—by experience. The computer estimating system manager must accept that the period of transition from manual to computerized estimating will involve a learning period for all concerned.

No matter how well an implementation plan has been prepared, or how much thought was given to procedures, unforeseen situations will arise. The computer estimating manager must have the flexibility to change the plan, standards, and procedures as needed. He or she must also be flexible enough to adapt the system over the long term as your company changes and grows. (You can also see why it's important to purchase an estimating system that offers the same kind of flexibility.)

Other Sources of Information

The best sources for implementation tips and guidelines are other home builders who have already introduced computer estimating into their companies. Ask the software company or the consultant with whom you are working for the names of companies in your area who use computer estimating. Unless the companies are in direct competition with you, most will be willing to exchange information.

Local user groups are another important resource, not only while you are investigating computer estimating, but also as part of your ongoing learning. Software companies will be able to give you the name of the user-group contact in your area.

Some vendors offer sample implementation plans, which can be a good starting point. But remember that builders have different requirements and different standards. A viable implementation plan must take these individual needs into account. Every company must approach implementation in its own way. A sample implementation plan, if available, will need to be customized to meet your needs. As mentioned before, it is important that you devote sufficient time to the preparation of a specific plan for your operation.

Ongoing training is another important resource. Find out whether the software vendor has a certified training program. Your local software representative may offer training classes as well.

Summary

Too often builders will invest in estimating software and never take it out of the box. These builders have failed to realize the importance of implementation. By following the guidelines outlined in this chapter, you can simplify the implementation process for your company and more quickly enjoy the efficiency computer estimating can bring.

Chapter Three

Advanced Tools

Basic computer estimating systems can significantly reduce the time it takes to prepare estimates. Depending on your needs, more advanced estimating tools are also available to further enhance your productivity and provide other useful capabilities. Advanced software packages may offer some or all of the following features.

Digitizer Software

If you do your own takeoffs—to determine lumber quantities, for example—digitizer software can be a handy tool. Many estimating systems designed for home builders offer digitizer software that interfaces with popular digitizer hardware. A digitizer is an electronic board with a stylus pen or puck pointer. The position of this pointer is transmitted directly to the computer. Using the digitizer with accompanying software, you can perform takeoffs directly from the plans—tracing and transferring dimensions to the computer estimating system. This greatly enhances takeoff efficiency. In fact, experience shows that a digitizer can double your takeoff productivity. Digitizing software typically takes areas, lengths, and counted points from your plans, graphically displays them onscreen, and allows you to instantly identify errors and omissions. (See Figure 3.1.)

While most estimating systems offer digitizing software, their digitizing capabilities may vary. For example, you may find a difference in scaling options and graphic capabilities between products. To help you evaluate your own digitizing needs, the following capabilities list gives you an idea of what is possible with good basic digitizer software:

- Counts objects and digitizes single lines, continuous lines, and areas.
- Automatically closes shapes when tracing areas.
- Displays a screen image of what the digitizer input looks like, making it easy to identify errors and omissions. For example, you can see on the computer screen an outline of the foundation as you trace over the plans. If you forget part of the foun-

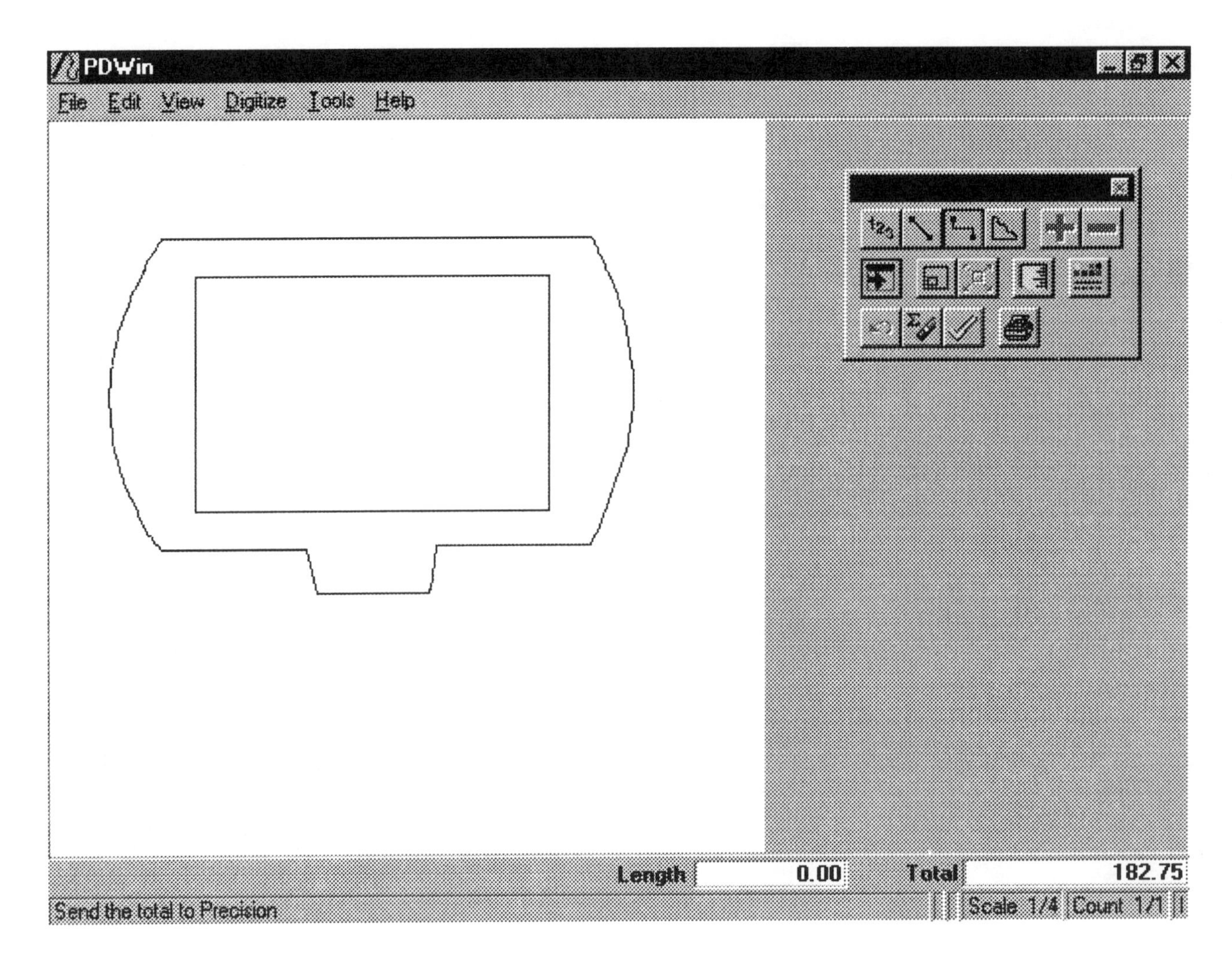

Figure 3.1 Sample Digitizer Screen

dation that separates the kitchen from the garage, you can see that, too.

- Prints a digitized image of what you have taken off from a home's plans.
- Provides a variety of scaling options. For example, autoscale is a very useful option when a plan has been copied and the original scale no longer applies. With autoscale, you can digitize a length, such as a 60-foot wall. The program then automatically computes the scale and uses it for the rest of the drawing.
- Runs the complete estimating system and digitizer from the digitizer tablet, thereby eliminating the need to use the keyboard or mouse.
- Labels a digitized length, area, or quantity for future use in the estimate.
- Is accurate up to 1/1000 inches.

Assemblies

When doing your own takeoffs, assemblies can be one of the most powerful aspects of your computer estimating system. (See Figure 3.2.) They allow you to easily generate all items and costs associated with a wall, foundation, or other building component simply by entering the component's dimensions. This can save a tremendous amount of time. To take off an 8-foot wall, for example, you just enter the wall's length. The wall assembly automatically generates the count and length of studs, plates, and blocking. Your assembly could even calculate takeoff quantities for drywall, insulation, sheathing, bracing, building paper, and exterior siding.

Assemblies can also prompt you with questions about building situations that may vary from job to job. A foundation assembly, for instance, may ask you whether or not you want forms, or whether you want to include sand in the foundation.

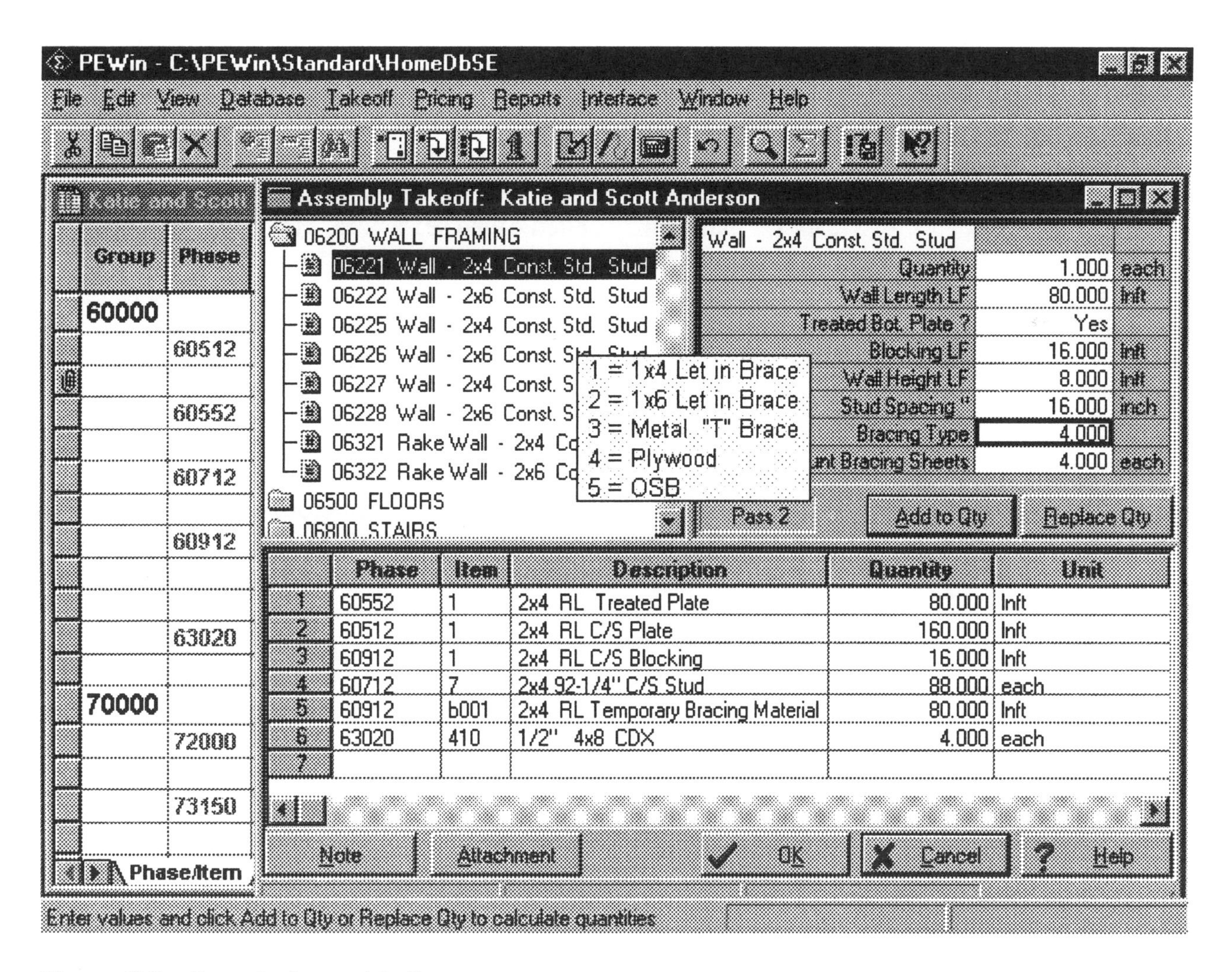

Figure 3.2 Sample Assembly Screen

The built-in logic available from "intelligent" assemblies can further increase the speed and accuracy of takeoff. If you change a parameter of a building component, the software automatically carries out preset instructions you have built into the assembly. Change a wall spec from 2x4 to 2x6, for example, and the assembly will automatically make the correct adjustments for the plates, blocking, studs, and insulation.

Intelligent assemblies allow you to build your own construction knowledge right into the estimating system. Essentially, you can build into your assemblies decisions that you would normally make during construction.

Some software comes with prebuilt assemblies as part of the database. These assemblies can be used as is or modified to fit your way of building. Always check out a prebuilt assembly before using it to determine if it needs modification. With most estimating software, you can also build your own unique assemblies from scratch.

Sequencing Capability

Some software lets you work with your estimate in any order you choose. (See Figure 3.3.) For example, you can view your estimate by material, trade contractor, or floor 1 and 2, and you can easily move back and forth between these views. These multiple ways to look at the estimate, often called sequences, can really save time. For example, you can look at your estimate by trade contractor to easily enter sub quotes. You can then instantly reorganize the budget into the different phases of construction, such as site work, concrete, and framing, to determine your material delivery schedule.

Estimate Management

Once you start creating your estimates electronically, a new issue arises. How do you organize and store numerous estimates so you can quickly retrieve them for future use? Some software packages are addressing this issue by providing archive and retrieval capabilities. With these systems, you can essentially create an electronic estimate library organized in any way you choose. For fast retrieval, some systems can sort your estimates in a variety of ways, such as by type of home, square footage, subdivision, person who prepared the estimate, and contract amount.

Archiving capabilities can do more than simply store your estimates. With some systems, you have access to reports for analyzing information across a number of estimates. For example, you could quickly see how your unit costs vary between jobs. You could also use these reports to identify what's working, and what's not, in your overall estimating process.

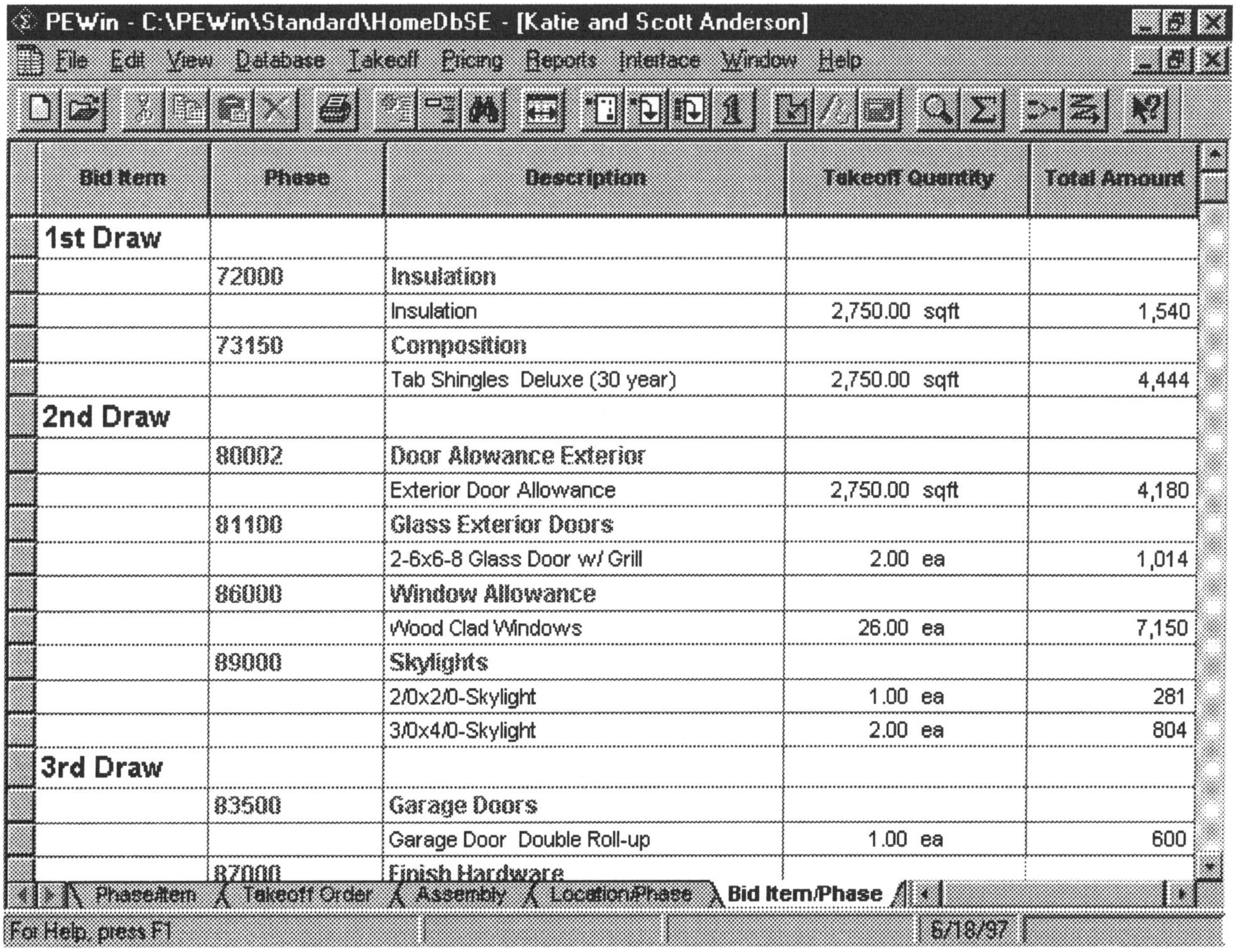

PEWin - C:\PEWin\Standard\HomeDbSE - [Katie and Scott Anderson]

File Edit View Database Takeoff Pricing Reports Interface Window Help

Bid Item	Phase	Description	Takeoff Quantity	Total Amount
1st Draw				
	72000	Insulation		
		Insulation	2,750.00 sqft	1,540
	73150	Composition		
		Tab Shingles Deluxe (30 year)	2,750.00 sqft	4,444
2nd Draw				
	80002	Door Alowance Exterior		
		Exterior Door Allowance	2,750.00 sqft	4,180
	81100	Glass Exterior Doors		
		2-6x6-8 Glass Door w/ Grill	2.00 ea	1,014
	86000	Window Allowance		
		Wood Clad Windows	26.00 ea	7,150
	89000	Skylights		
		2/0x2/0-Skylight	1.00 ea	281
		3/0x4/0-Skylight	2.00 ea	804
3rd Draw				
	83500	Garage Doors		
		Garage Door Double Roll-up	1.00 ea	600
	87000	Finish Hardware		

Phase/Item | Takeoff Order | Assembly | Location/Phase | **Bid Item/Phase**

For Help, press F1 | 6/18/97

Figure 3.3 Sample Sequencing Capabilities

Future Trends

When it comes to computer technology, one thing is certain: there will always be a new and better way to do things. Estimating software is no exception. The estimating advances you've just read about in this chapter will soon step aside to make room for more exciting breakthroughs.

It's not difficult to predict that many of the upcoming trends and changes will center around the Internet. On-line pricing is one such trend. Some homebuilding suppliers are already providing free price lists in electronic form. The logical next step is to provide up-to-the-minute pricing on their web sites. With on-line pricing it could take literally seconds to keep your database up to date.

Tomorrow's estimating software will also be more "object-oriented." For example, instead of viewing just lists of text and numbers, you could point to a door, wall, or other object on your home plan and the estimating system will give you the supplier, price, notes, a picture, and sound bite describing the item and any other information connected to that object. These capabilities will most likely become available through the tighter integration of estimating and CAD systems. Current integration capabilities are discussed in Chapter 5.

Overall, you'll find that it will be much easier to pass information among the construction software programs of the future. This will only lead to better communication within your company and with outside architects, suppliers, your banker, and anyone else you do business with. Will it lead to a paperless office? Probably not, but it will significantly reduce your paper volume, as well as the chance for errors, double entry, and redundancy.

Chapter Four

Preparing Reports and Other Documents

A major advantage of computer estimating is the ability to quickly and easily prepare useful reports and documents. The computer can take the numbers contained in an estimate and rearrange them to get a wide variety of reports or summaries. You can prepare and print such items as quote sheets, purchase orders, spec sheets for your bank, and presentation-quality reports for clients.

Reports as a Marketing Tool

Now more than ever, home buyers demand quality—not only in the way you build their homes but also in the way you run your business. Consequently, the professional image you present to clients has become an important component in obtaining work and gaining future referrals.

The reporting options available with today's estimating systems are designed to demonstrate your professionalism to clients through the quality of information you can provide. Whether your clients are requesting a preliminary estimate or want to see a breakdown of costs, you can quickly create a report to answer their questions.

Not only can you provide information quickly, you can present it with a professional polish. For example, you can use different type styles and sizes, make some items bold or italic, and even use color to highlight a particular part of your report. For even fancier graphics, such as pie charts and graphs, many systems export data to applications such as Excel or Lotus.

The ability to dress-up your reports is still a fairly new capability for most estimating systems. Many builders have found this capability to be a definite business advantage. When looking for an estimating system, make sure it offers reporting capabilities to support the professional image you want to present.

Standard and Custom Reports and Documents

Most estimating systems provide a standard set of commonly used reports and documents for your immediate use. A quote sheet, for example, may be available for you to send out to suppliers and trade contractors. These documents can usually be modified to fit your company's individual needs.

But what about that custom report your company has developed—the one that's the hallmark of your operation and sets you apart from your competition? To accommodate this need, many systems offer custom report-writing capabilities. With a custom report writer, you can design a report and use it again for successive projects. You can also create reports to answer specific questions you may have. For instance, maybe you'd like a breakdown of cost by each room in the house.

Software packages vary in their custom-report-generating ability and approach. Some products produce custom reports quickly by stepping you through a few simple question prompts, such as the range of cost codes you'd like to see. Others offer more advanced features, such as free-form report design and sophisticated word-processing capabilities.

Sample Reports

What follows is a sampling of reports and documents that can be produced, depending on the particular software package you use. Keep in mind that not all builders will need each of these reports. Identify the reporting needs of your company and select a package that addresses them. As your business grows and the industry changes, you may find that your reporting needs will also change. So you may want to consider software that offers flexible reporting features.

Bill of Materials

Once you've created an estimate, you need to start the process of purchasing materials. A bill of materials sorts estimate items and quantities by material type, making it easier to see what materials need to be purchased. (See Figure 4.1.)

Quote Sheets

A quote sheet contains the same information as your bill of materials but is sent to trade contractors and suppliers to streamline the request for and submission of prices. (See Figure 4.2.) All the recipient has to do is enter prices.

Bid Comparison Report

Once you receive your trade contractor and supplier bids, you have to select the best price. Many estimating systems offer a bid comparison report so you can quickly determine which bid to accept for individual work items. (See Figure 4.3.)

Field Reports

As the job begins, so does the communication to on-site supervisors and project managers. The type of field reports you use depends largely on what your onsite managers need to see. The sample field report shown in this chapter, which indicates where material will be used on the job, is only one of many field reports that can be generated with an estimating system. (See Figure 4.4.)

Audit Trail

Sometimes you need to look back to verify that you added an item to the estimate or to determine why you made a certain estimating decision. An audit trail provides a history of how an estimate was put together. This report may show a detailed list of each estimate item, including when it was added, any changes made, who made the changes, and attached notes. (See Figure 4.5.)

Estimate Comparison Report

Many estimating systems also give you the capability to check your current estimate numbers against previous jobs. Comparison reports enable you to analyze information, such as unit-cost variances, across a number of estimates. (See Figure 4.6.)

Purchase Orders and Contracts

Throughout the job, purchase orders and contracts can be sent to suppliers and trade contractors to authorize work and the delivery of materials. In some systems, bill of materials information is sent to the accounting software that prepares and issues purchase orders. (See Figure 4.7.)

Client Reports

Last, but definitely not least, what do you provide to clients when they ask for a bid, need to see their budget, or question a cost? It depends on what you would like to provide to them. The variety of client reports is endless, and the type of report you use depends on the type of homes you build, the typical questions your clients ask, and how your company wants to present client budget information. (See Figure 4.8.)

Summary

How you communicate with your clients, as well as your field staff, banker, trade contractors, and suppliers is a key component to your success as a home builder. Estimating software can help you communicate accurately and professionally with impressive-looking reports that allow you to provide information in a variety of ways to a variety of audiences.

Coffman Construction **Bill of Materials** ***Smith Home*** ***10/24/97 Page 1*** ***9:47 AM***

Description	Takeoff Qty		Order Qty		Unit Price	Amount
Appliances						
Kenmore Washer	1.00	each	1.00	each	475.00	475.00
Kenmore Dryer	1.00	each	1.00	each	260.00	260.00
GE Range	1.00	each	1.00	each	600.00	600.00
GE Refrigerator	1.00	each	1.00	each	960.00	960.00
						2,295.00
Concrete						
Sand Base	12.00	cuyd	15.00	ton	10.00	150.00
Slab Concrete 3000 psi	16.00	cuyd	16.00	cuyd	45.00	720.00
6x6 1010 Welded Wire Mesh	997.00	sqft	997.00	sqft	0.10	99.70
6 ml Vapor Barrier	997.00	sqft	1.00	roll	18.00	18.00
						987.70
Doors & Windows						
Standard Single Leaf Door	9.00	each	9.00	each	150.00	1,350.00
Sliding Double Door	4.00	each	4.00	each	325.00	1,300.00
Pocket Door	1.00	each	1.00	each	275.00	275.00
Garage Door	1.00	each	1.00	each	1,500.00	1,500.00
Double Bi-Fold	1.00	each	1.00	each	250.00	250.00
						4,675.00
Framing Material						
2x4 RL C/S	810.44	lnft	0.540	mbf	677.14	365.85
2x4 RL Treated	426.64	lnft	0.284	mbf	677.004	192.56
2x4 92-1/4" C/S	339.00	each	1.808	mbf	677.03	1,224.07
2x4 8 C/S	4.00	each	0.021	mbf	678.00	14.46
2x4 10 C/S	64.00	each	0.427	mbf	677.00	288.85
2x4 14 C/S	20.00	each	0.187	mbf	676.80	126.34
2x4 16 C/S	8.00	each	0.085	mbf	378.00	32.26
2x6 8 #1	39.00	each	0.312	mbf	444.00	138.53
2x4 RL C/S	759.27	lnft	0.506	mbf	677.16	342.76
2x4 RL C/S	147.35	lnft	0.098	mbf	676.995	66.50
4x8 RL #1	112.00	lnft	0.299	mbf	700.00	209.07
4x12 RL #1	96.83	lnft	0.387	mbf	628.00	243.24
						3,244.49
Framing OSB/Plywood						
1/2" 4x8 OSB	48.00	each	1.536	mbf	245.00	376.32
						376.32
Gypsum Board						
1/2" X Drywall Walls	2,528.00	sqft	2,528.00	sqft	0.20	505.60
1/2" Drywall Walls	2,573.00	sqft	2,573.00	sqft	0.50	1,286.50
						1,792.10
Insulation						
5/8" 4x8 Thermax	91.00	each	2,912.00	sqft	8.00	23,296.00
5/8" 4x8 Thermax	2,573.00	sqft	2,573.00	sqft	0.20	514.60
						23,810.60
Interior Trim						
Colonial Base FJ 3-1/2"	594.00	lnft	594.00	lnft	0.550	326.70
						326.70

Figure 4.1 Sample Bill of Materials Report

Coffman Construction			**Bill of Materials** *Smith Home*				*10/24/97* *9:47 AM* *Page 2*
Description	**Takeoff Qty**		**Order Qty**		**Unit Price**	**Amount**	
Plumbing Fixtures							
Standard Bathroom Sink	1.00	each	1.00	each	125.00	125.00	
Standard Bath Tub	1.00	each	1.00	each	560.00	560.00	
Standard Toilet	2.00	each	2.00	each	125.00	250.00	
						935.00	
Roof Trusses							
20' Truss	4.00	each	4.00	each	120.00	480.00	
24' Truss	29.00	each	29.00	each	130.00	3,770.00	
						4,250.00	
Siding							
1x6 RL Cedar Lap Resawn	2,573.00	sqft	2.00	mbf	1,045.00	2,090.00	
						2,090.00	
Windows - Vinyl							
3/0 x 3/0 Vinyl Window	14.00	each	14.00	each	70.00	980.00	
						980.00	
Grand Total						***45,762.91***	

Figure 4.1 Sample Bill of Materials Report *(continued)*

Coffman Construction | **Bill of Materials Quote Sheet** *Smith Home* | *10/24/97 Page 1* *9:53 AM*

Description	Takeoff Qty		Order Qty		Unit Price
Appliances					
Kenmore Washer	1.00	each	1.00	each	________
Kenmore Dryer	1.00	each	1.00	each	________
GE Range	1.00	each	1.00	each	________
GE Refrigerator	1.00	each	1.00	each	________
Concrete					
Sand Base	12.00	cuyd	15.00	ton	________
Slab Concrete 3000 psi	16.00	cuyd	16.00	cuyd	________
6x6 1010 Welded Wire Mesh	997.00	sqft	997.00	sqft	________
6 ml Vapor Barrier	997.00	sqft	1.00	roll	________
Doors & Windows					
Standard Single Leaf Door	9.00	each	9.00	each	________
Sliding Double Door	4.00	each	4.00	each	________
Pocket Door	1.00	each	1.00	each	________
Garage Door	1.00	each	1.00	each	________
Double Bi-Fold	1.00	each	1.00	each	________
Framing Material					
2x4 RL C/S	810.44	lnft	0.540	mbf	________
2x4 RL Treated	426.64	lnft	0.284	mbf	________
2x4 92-1/4" C/S	339.00	each	1.808	mbf	________
2x4 8 C/S	4.00	each	0.021	mbf	________
2x4 10 C/S	64.00	each	0.427	mbf	________
2x4 14 C/S	20.00	each	0.187	mbf	________
2x4 16 C/S	8.00	each	0.085	mbf	________
2x6 8 #1	39.00	each	0.312	mbf	________
2x4 RL C/S	759.27	lnft	0.506	mbf	________
2x4 RL C/S	147.35	lnft	0.098	mbf	________
4x8 RL #1	112.00	lnft	0.299	mbf	________
4x12 RL #1	96.83	lnft	0.387	mbf	________
Framing OSB/Plywood					
1/2" 4x8 OSB	48.00	each	1.536	mbf	________
Gypsum Board					
1/2" X Drywall Walls	2,528.00	sqft	2,528.00	sqft	________
1/2" Drywall Walls	2,573.00	sqft	2,573.00	sqft	________
Insulation					
5/8" 4x8 Thermax	91.00	each	2,912.00	sqft	________
5/8" 4x8 Thermax	2,573.00	sqft	2,573.00	sqft	________
Plumbing Fixtures					
Standard Bathroom Sink	1.00	each	1.00	each	________
Standard Bath Tub	1.00	each	1.00	each	________
Standard Toilet	2.00	each	2.00	each	________
Roof Trusses					
20' Truss	4.00	each	4.00	each	________
24' Truss	29.00	each	29.00	each	________
Siding					
1x6 RL Cedar Lap Resawn	2,573.00	sqft	2.00	mbf	________
Windows - Vinyl					
3/0 x 3/0 Vinyl Window	14.00	each	14.00	each	________

Figure 4.2 Sample Quote Sheet

Coffman Construction

Vendor Comparison Report
Smith Home
Mat Quote Sheet ID SM01

Item Description	Budgeted Amount	Divers Doors and Windows	Columbia Exteriors Inc.
16x7 Redwood Roll-up Garage Door	800.00	675.00	799.00
3068 15 Lite Dual Glazed Glass Panel Door	441.00	449.00	450.00
6068 1 Lite Dual Glazed Wood/Glass French D	215.00	199.00	220.00
7068 1 Lite Dual Glazed w/Grill Patio Unit	1267.00	1150.00	1200.00
3068 Oak 6-Panel Wood Door	520.00	479.00	529.00
4020 Skylight	327.00	300.00	250.00
5050 Wood Clad DBL Casement w/Screens	895.00	750.00	700.00
Totals	4465.00	4002.00	4148.00
Variance		463.00	317.00

Figure 4.3 Sample Bid Comparison Report

Coffman Construction		Field Report Smith Home		10/24/97 Page 1 10:05 AM

Item	Description	Takeoff Qty		Labor				Material	
				Productivity		Quantity		Quantity	
60000	**LUMBER & FRAMING**								
60200	**Framing Labor**								
15	Layout & Detail	3.00	sqft			3.00	sqft		
20	Plate Labor	3.00	sqft			3.00	sqft		
70	Plumb & Line Walls	3.00	sqft			3.00	sqft		
90	Joisting Floor System	3.00	sqft			3.00	sqft		
60411	**Floor Joist #1**								
108	2x6x8 #1 Joist	39.00	each					0.312	mbf
60512	**Plates #2**								
1	2x4 RL C/S Plate	810.44	lnft					0.540	mbf
60552	**Treated Plates**								
1	2x4 RL Treated Plate	426.64	lnft					0.284	mbf
60712	**Studs C/S & #2**								
7	2x4 92-1/4" C/S Stud	339.00	each	12.00000	each/hour	28.25	hour	1.808	mbf
8	2x4x8 C/S Stud	4.00	each	14.00000	each/hour	0.286	hour	0.021	mbf
10	2x4x10 C/S Stud	64.00	each	14.00000	each/hour	4.571	hour	0.427	mbf
14	2x4x14 C/S Stud	20.00	each	17.00000	each/hour	1.176	hour	0.187	mbf
16	2x4x16 C/S Stud	8.00	each			8.00	each	0.085	mbf
	Studs C/S & #2 34.284 Labor hours								
60912	**Blocking C/S & #2**								
1	2x4 RL C/S Blocking	759.27	lnft					0.506	mbf
b001	2x4 RL Bracing Material	147.35	lnft					0.098	mbf
61011	**Headers & Beams #1**								
4201	4x8 RL #1 Headers & Beams	112.00	lnft					0.299	mbf
4401	4x12 RL #1 Headers & Beams	96.83	lnft					0.387	mbf
62011	**Rafters #1**								
01	20' Truss	4.00	each					4.00	each
03	24' Truss	29.00	each	2.00000	each/hour	14.50	hour	29.00	each
	Rafters #1 14.50 Labor hours								
63020	**Sheathing**								
430	1/2" 4x8 OSB	48.00	each	5.00000	each/hour	9.60	hour	1.536	mbf
580	5/8" 4x8 Thermax	91.00	each					2,912.00	sqft
	Sheathing 9.60 Labor hours								
67010	**Wood Siding**								
205	1x6 RL Cedar Lap Resawn	2,573.00	sqft	10.00000	sqft/hour	257.30	hour	1.287	mbf
	Wood Siding 257.30 Labor hours								
	LUMBER & FRAMING 315.684 Labor hours								

Figure 4.4 Sample Field Report

Coffman Construction **Takeoff Audit Report** *10/24/97 Page 1*
Smith Home *9:45 AM*

Date	Time	Sequence	Phase	Item	Description	Takeoff Qty		Location
8/5/97	8:15 AM	6,624	10200	10	Building Permits	2,800.00	sqft	
8/5/97	8:15 AM	6,625	10200	30	Builders Risk Insurance	1.00	each	
8/5/97	8:15 AM	6,627	10250	50	Drawings and Specifications	1.00	each	
8/5/97	8:15 AM	6,628	10250	60	Surveying	1.00	each	
8/5/97	8:15 AM	6,636	10300	10	Electrical Service	4.00	mo	
8/5/97	8:15 AM	6,637	10300	20	Electrical Hookup	1.00	each	
8/5/97	8:15 AM	6,634	10300	30	Water Service	4.00	mo	
8/5/97	8:15 AM	6,635	10300	40	Water Hookup	1.00	each	
8/5/97	8:15 AM	6,632	10700	20	Final Cleanup	2,800.00	sqft	
8/5/97	8:28 AM	6,996	81050	50	Garage Door	1.00	each	Garage
8/5/97	12:17 PM	676	30100	s104	Building Slab Labor	997.00	sqft	House
8/5/97	12:18 PM	675	30700	6	6 ml Vapor Barrier	997.00	sqft	House
8/5/97	12:18 PM	672	30700	10	Sand Base	12.00	cuyd	House
8/5/97	12:18 PM	674	32500	1010	6x6 1010 Welded Wire Mesh	997.00	sqft	House
8/5/97	12:18 PM	673	35000	30	Slab Concrete 3000 psi	16.00	cuyd	House
8/5/97	12:21 PM	1,113	67010	205	1x6 RL Cedar Lap Resawn	2,573.00	sqft	House
8/5/97	12:22 PM	1,114	72000	111	3-1/2" R11 Faced Batt	2,573.00	sqft	House
8/5/97	12:22 PM	979	63020	430	1/2" 4x8 OSB	48.00	each	House
8/5/97	12:22 PM	1,110	63020	580	5/8" 4x8 Thermax	91.00	each	House
8/5/97	12:23 PM	1,954	61011	4201	4x8 RL #1 Headers & Beams	112.00	lnft	House
8/5/97	12:23 PM	1,648	61011	4401	4x12 RL #1 Headers & Beams	96.83	lnft	House
8/5/97	12:24 PM	981	62011	03	24' Truss	29.00	each	House
8/5/97	12:24 PM	980	73150	20	Tab Shingles Deluxe (25 year)	1,476.19	sqft	House
8/5/97	12:25 PM	1,647	81050	30	Standard Single Leaf Door	9.00	each	House
8/5/97	12:25 PM	2,208	81050	40	Sliding Double Door	4.00	each	House
8/5/97	12:25 PM	6,792	86240	10	3/0 x 3/0 Vinyl Window	4.00	each	Garage
8/5/97	12:26 PM	3,483	86240	11	Anderson 3/0 x 3/6 Vinyl Window	2.00	each	House
8/5/97	12:26 PM	1,953	86240	14	3/0 x 5/0 Vinyl Window	8.00	each	House
8/5/97	12:26 PM	832	92500	400	1/2" X Drywall Walls	2,528.00	sqft	House
8/5/97	12:27 PM	1,115	92500	401	1/2" Drywall Walls	2,573.00	sqft	House
8/5/97	12:29 PM	826	60552	1	2x4 RL Treated Plate	426.64	lnft	House
8/5/97	12:29 PM	828	60712	7	2x4 92-1/4" C/S Stud	339.00	each	House
8/5/97	12:30 PM	1,100	60712	8	2x4x8 C/S Stud	4.00	each	House
8/5/97	12:30 PM	1,101	60712	10	2x4x10 C/S Stud	64.00	each	House
8/5/97	12:31 PM	1,103	60712	14	2x4x14 C/S Stud	20.00	each	House
8/5/97	12:31 PM	7,118	60712	16	2x4x16 C/S Stud	8.00	each	Garage
8/5/97	12:32 PM	829	60912	1	2x4 RL C/S Blocking	759.27	lnft	House
8/5/97	12:32 PM	831	60912	b001	2x4 RL Bracing Material	147.35	lnft	House
8/5/97	12:34 PM	833	94020	220	Colonial Base FJ 3-1/2"	594.00	lnft	House
8/5/97	12:34 PM	834	99000	10	Painting Interior Wallboard	5,101.00	sqft	House
8/5/97	12:35 PM	1,118	99010	10	Painting Exterior Walls	2,573.00	sqft	House
8/5/97	12:35 PM	1,339	110110	110	Oak Base Recessed Plywd Door	21.35	lnft	House
8/5/97	12:36 PM	1,340	110110	120	Oak Upper Recessed Plywd Door	21.35	lnft	House
8/5/97	12:36 PM	1,596	154050	20	Standard Toilet	2.00	each	House
8/5/97	12:45 PM	7,352	81050	10	Double Bi-Fold	1.00	each	
8/5/97	1:30 PM	5,839	60411	108	2x6x8 #1 Joist	39.00	each	House
8/5/97	1:31 PM	5,849	60200	70	Plumb & Line Walls	3.00	sqft	

Figure 4.5 Sample Audit Report

Estimate Comparison Report

8/6/97 8:47:40 PM

DESCRIPTION	Est. Name	Bid Date	Job Size		Unit Cost		Bid Amount
Project Type: **Harbor Place**							
270 Channel Drive Seattle, WA	Fairfield	4/17/97	2,900 sqft		89.66/sqft		260,000
440 Pond Way Seattle, WA	Stanford	7/18/97	2,500 sqft		75.20/sqft		188,000
410 Pond Way Seattle, WA	Williamsburg	7/10/97	3,100 sqft		77.42/sqft		240,000
129 Channel Drive Seattle, WA	Fairfield	7/22/97	5,200 sqft		115.38/sqft		600,000
442 River Road	Southern	8/17/97	4,200 sqft		121.43/sqft		510,000
		Total	17,900 sqft	***Average***	100.45/sqft	**$**	**1,798,000**
Project Type: **Hillcrest**							
3400 Mountain View Road Portland, OR	Fairfield	3/13/97	2,900 sqft		88.07/sqft		255,400
4289 Hillcrest Drive Portland, OR	Georgia	6/10/97	3,500 sqft		97.14/sqft		340,000
3290 Mountain View Road Portland, OR	Williamsburg	6/11/97	3,000 sqft		96.63/sqft		289,900
1240 Walton Way Portland, OR	Southern	8/14/97	4,100 sqft		147.32/sqft		604,000
8341 Hillcrest Drive	Stanford	9/10/97	2,200 sqft		113.64/sqft		250,000
		Total	15,700 sqft	***Average***	110.78/sqft	**$**	**1,739,300**

Figure 4.6 Sample Estimate Comparison Report

PURCHASE ORDER

From:
Coffman Construction

FAX:

Job: Steward Home
QS ID: 44 PO# 10013

Date: 8/6/97

To:
Sanderson Doors
4500 N. Perkins Road
Portland, OR 97220

FAX:

Ship to:
Steward Home
19226 S. Dillon Avenue
Portland, OR 97002

Ord by:______ Deliver:______ Terms:______ Ship Via:______ FOB:______

Item	Quantity	Unit	Description	Unit Price	Extension
1	3	each	Standard Single Leaf Door	425.0000	1,275.00
2	1	each	Double Bi - Fold	225.0000	225.00
3	3	each	Sliding Double Door	950.0000	2,850.00
4	4	each	Pocket Door	250.0000	1,000.00
5	1	each	Garage Door	1475.0000	1,475.00
				Total: ($)	6,825.00

Figure 4.7 Sample Purchase Order

Coffman Construction **Spreadsheet Report** *10/24/97* *Page 1*
Smith Home *9:42 AM*

Group	Phase	Description	Takeoff Quantity	Grand Total
10000		**GENERAL CONDITIONS**		
	10200	**Permits & Insurance**		
		Building Permits	2,800.00 sqft	4,982
		Builders Risk Insurance	1.00 each	297
	10250	**Architect**		
		Drawings and Specifications	1.00 each	1,423
		Surveying	1.00 each	356
	10300	**Utilities**		
		Electrical Service	4.00 mo	308
		Electrical Hookup	1.00 each	83
		Water Service	4.00 mo	95
		Water Hookup	1.00 each	534
	10500	**Rented Equipment**		
		Portable Toilets	1.00 each	237
	10700	**Cleanup**		
		Final Cleanup	2,800.00 sqft	498
30000		**FOUNDATIONS & SLABS**		
	30100	**Foundation Labor & Equip**		
		Building Slab Labor	997.00 sqft	852
	30700	**Base Matl, Waterproofing**		
		6 ml Vapor Barrier	997.00 sqft	21
		Sand Base	12.00 cuyd	178
	32500	**Mesh Reinforcing**		
		6x6 1010 Welded Wire Mesh	997.00 sqft	118
	35000	**Slabs**		
		Slab Concrete 3000 psi	16.00 cuyd	12,242
60000		**LUMBER & FRAMING**		
	60200	**Framing Labor**		
		Layout & Detail	3.00 sqft	0
		Plate Labor	3.00 sqft	0
		Plumb & Line Walls	3.00 sqft	0
		Joisting Floor System	3.00 sqft	1
	60411	**Floor Joist #1**		
		2x6x8 #1 Joist	39.00 each	164
	60512	**Plates #2**		
		2x4 RL C/S Plate	810.44 lnft	434
	60552	**Treated Plates**		
		2x4 RL Treated Plate	426.64 lnft	228
	60712	**Studs C/S & #2**		
		2x4 92-1/4" C/S Stud	339.00 each	1,955
		2x4x8 C/S Stud	4.00 each	22
		2x4x10 C/S Stud	64.00 each	424
		2x4x14 C/S Stud	20.00 each	171
		2x4x16 C/S Stud	8.00 each	41
	60912	**Blocking C/S & #2**		
		2x4 RL C/S Blocking	759.27 lnft	407
		2x4 RL Bracing Material	147.35 lnft	79
	61011	**Headers & Beams #1**		
		4x8 RL #1 Headers & Beams	112.00 lnft	248
		4x12 RL #1 Headers & Beams	96.83 lnft	289
	62011	**Rafters #1**		
		20' Truss	4.00 each	569

Figure 4.8 Sample Client Report

Coffman Construction **Spreadsheet Report** *10/24/97 Page 2*
Smith Home *9:42 AM*

Group	Phase	Description	Takeoff Quantity	Grand Total
	62011	**Rafters #1**		
		24' Truss	29.00 each	4,730
	63020	**Sheathing**		
		1/2" 4x8 OSB	48.00 each	617
		5/8" 4x8 Thermax	91.00 each	27,634
	67010	**Wood Siding**		
		1x6 RL Cedar Lap Resawn	2,573.00 sqft	6,173
70000		**ROOF/GUTTERS/INSUL/VENT**		
	72000	**Insulation**		
		3-1/2" R11 Faced Batt	2,573.00 sqft	610
	73150	**Roofing - Composition**		
		Tab Shingles Deluxe (25 year)	1,476.19 sqft	1,488
80000		**DOORS AND WINDOWS**		
	81050	**Doors**		
		Double Bi-Fold	1.00 each	297
		Pocket Door	1.00 each	326
		Standard Single Leaf Door	9.00 each	1,922
		Sliding Double Door	4.00 each	1,542
		Garage Door	1.00 each	1,922
	86240	**Vinyl Windows**		
		3/0 x 3/0 Vinyl Window	4.00 each	474
		Anderson 3/0 x 3/6 Vinyl Window	2.00 each	166
		3/0 x 5/0 Vinyl Window	8.00 each	664
90000		**DRYWALL/TRIM/PAINT/CARP**		
	92500	**Drywall**		
		1/2" X Drywall Walls	2,528.00 sqft	2,099
		1/2" Drywall Walls	2,573.00 sqft	1,526
	94020	**Base**		
		Colonial Base FJ 3-1/2"	594.00 lnft	493
	99000	**Painting**		
		Painting Interior Wallboard	5,101.00 sqft	1,150
	99010	**Painting Options**		
		Painting Exterior Walls	2,573.00 sqft	458
110000		**CABINETS & APPLIANCES**		
	110110	**Factory Hardwood Cabinets**		
		Oak Base Recessed Plywd Door	21.35 lnft	1,266
		Oak Upper Recessed Plywd Door	21.35 lnft	1,266
	110500	**Appliances**		
		GE Range	1.00 each	712
		GE Refrigerator	1.00 each	1,139
		Kenmore Washer	1.00 each	563
		Kenmore Dryer	1.00 each	308
150000		**MECHANICAL**		
	154050	**Plumbing Fixtures**		
		Standard Bath Tub	1.00 each	902
		Standard Toilet	2.00 each	652
		Standard Bathroom Sink	1.00 each	326

Figure 4.8 Sample Client Report *(continued)*

Coffman Construction	**Spreadsheet Report** ***Smith Home***	***10/24/97 Page 3A*** ***9:42 AM***

Estimate Totals

Labor	8,939		446.514 hrs	
Material	45,017			
Subcontract	15,496			
Other	5,310			
	74,762	**74,762**		
Bond	3,279			B
Office Overhead	1,951		2.500 %	T
Lot Cost	60,000			L
Profit & Overhead	11,214		15.000 %	C
Insurance Cost	756		5.000 $ /	1,000.000 T
	Total	***151,962***		

Figure 4.8 Sample Client Report *(continued)*

Chapter Five

Linking Computer Estimating with Other Software

Estimating is certainly not the only aspect of home building that can benefit from computer applications. Home builders can achieve considerable efficiency through computer-based job-cost accounting, purchasing, scheduling, pricing, and computer-aided design (CAD). Because of the close relationship between these functions, users of computer estimating systems are increasingly interested in integrating them. Linking the functions electronically can eliminate repetitive data entry and offer considerable time savings. (See Figure 5.1.)

Although integration can enhance productivity, home builders should not necessarily try it immediately. You first need to become familiar with each application individually *before* you try to integrate them. If your company is larger, you will also need to address internal issues concerning access to and sharing of informa-

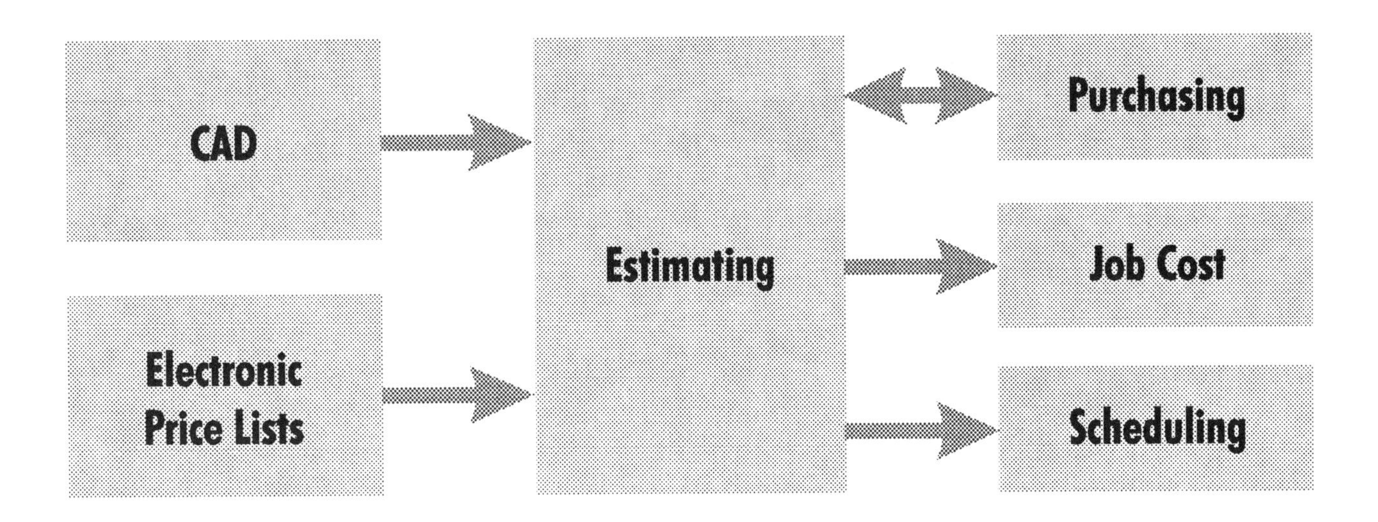

Figure 5.1 Integration Flow Chart

tion between the different parts of your operation. For integration to work, all staff members affected must function together as a team. If you want to integrate several functions, the best approach is to do it in stages.

If you are interested in integration—even if it's in the future—read this chapter thoroughly. Then think about what functions you might want to link. This information could influence your choice of estimating or other software. Ask the software company questions about existing integration capabilities and plans for future integration between products. Planning ahead can be an important step toward achieving successful integration.

Understanding Integration

Integrated systems work by sharing data among two or more programs. For example, information can flow from an estimating program to an accounting, purchasing, or scheduling program, or from a CAD system to an estimating program. Each program translates the information into a format usable by another complementary application.

There are two basic approaches to integration. You can purchase a system from a single software company that offers built-in links for different functions. For example, some CAD systems offer a few basic estimating capabilities.

Another approach is to buy software that links multiple applications from different vendors. Such an approach gives you the flexibility to purchase estimating software from a company with expertise in that area and other software, such as scheduling, from a company that is an expert in that field. (It is not uncommon for software companies to offer both built-in links as well as integration with software from other companies.) After the initial setup is in place, using an integrated system is fairly straightforward and brings numerous benefits. Here is an overview of several functions you might want to consider integrating and an explanation of how such integration could streamline your operation.

Job Cost Accounting

The most common and usually the first link made is between computer estimating and job cost accounting systems. Essentially, by integrating these two systems, you can gain a better understanding of your true costs and how they deviate from your budget. Integration makes this cost-to-budget comparison much more possible because you don't have to spend hours manually entering costs from your final budget into your job cost system. It's all done automatically and with less chance for error.

Several software companies offer accounting modules that link directly with their own estimating software. In some cases, you also can link software products offered by different vendors.

In either case, cost tracking can be done according to NAHB cost codes or your own chart of accounts.

Purchasing Process

Turning an estimate into a set of purchase orders is time-consuming and yet another opportunity for mistakes to occur. That's what makes purchasing software such a logical extension of your computer estimating system. Applications are available that handle the entire purchasing process, including the following:

- Generating a series of quote sheets to send to vendors.
- Comparing quotes when you receive them.
- Tracking the cost difference between what has been purchased and what was estimated. For example, you could see that you are $1,000 under budget for concrete and lumber, with 75 percent of the job materials still to be purchased.
- Breaking down the estimate by vendor and trade contractor and issuing purchase orders.
- Providing instant access to estimate totals whenever you need them.

In some cases, you can even transfer data from purchasing back to the estimate. For example, if you receive a price quote before your estimate is final, you can use this information to update your estimate.

Some applications allow further integration between purchasing and scheduling by using the project schedule to trigger the release of purchase orders and trade contracts. Should the schedule change, purchase-order dates are automatically adjusted to reflect the revised schedule.

And some software makes sure a close tie exists between purchasing and job cost. For instance, if a purchase order exceeds the job cost budget for that particular cost item, then a warning displays to indicate cost overruns before they actually occur.

Scheduling

If you build custom homes, you may want to provide clients with a simple schedule. If you're working on a subdivision, you may want to schedule activities so that, for example, all work is done sequentially to keep your trade contractors continually working.

Interfaces are available that permit you to link data between estimating and project-scheduling packages. With such a link, you can automatically create a planning base of tasks or activities using material items, notes, and construction assumptions from your estimate. Consequently, you can assign tasks in ways that better represent the estimate. Complete projected costs and resource details are available before the job is started, instead of piecemeal information acquired after construction has begun. Such integration also allows you to update a schedule at the same time that you revise an estimate.

Pricing

If you store prices in your database, you can use integration software to help you keep those prices up to date. Some building suppliers offer electronic price lists. A pricing application can automatically transfer the most current prices from suppliers into your estimating database.

Pricing integration is a fast and simple way to ensure that your prices stay current. It also eliminates errors caused by manually entering new prices into the database.

Computer-Aided Design (CAD)

As a builder, you know how costs are driving the designs of today's homes. Customers come to you with budgets and design ideas. Your clients want vaulted ceilings, hot tubs, and other customized specialty features. And you have to find ways to provide those options without pricing your homes out of the market.

Linking CAD and computer estimating can help you better serve customers who want both quality and value. If the owners can't decide whether to expand a room, you can show them what it will look like and how it will affect their budget. If market studies show that local consumers want 9-foot ceilings, you can do a "what if" analysis to determine how you can best offer this feature. The key to any CAD-to-estimating integration is that you can quickly create a variety of design-cost alternatives, which can become a valuable marketing and customer-service tool.

Software products have been available for several years that transfer takeoff dimensions and specifications directly from a CAD program to an estimating package. Recently, CAD-to-estimate integration has taken a step forward with the advent of *object-based* CAD systems. Basically, this advance means that home designers using CAD no longer have to create drawings with geometric shapes, such as lines, circles, and arcs. Instead, their drawings can consist of more realistic objects such as doors, windows, walls, and floors. These objects can contain specifications, costs, dimensions, a supplier's name, and other relevant information. In other words, object-based CAD more practically replicates the design and construction process.

An interface between object-based CAD and an estimating system is much easier to use because you can obtain more budget information directly from the CAD drawing. For example, if you take off a concrete slab with openings for stairwells, the slab (object) will report its area minus the openings.

With a few object-based systems, you can continue to refine and modify a budget from the CAD drawing after takeoff has occurred. Some interfaces also automatically update the budget whenever a change is made to the drawing.

Pulling It All Together

Building a home involves many different but related tasks—whether you are designing the home, estimating its costs, purchasing materials and services, scheduling activities, or controlling actual expenses. Integrated software can help you pull these tasks together, eliminating inefficiencies and streamlining your overall workflow.

Chapter Six

Home Builders Using Estimating Software

Some of the best advice on how to select, implement, and use computer estimating can come from other builders. In fact, survey data from the NAHB International Builders Show (1997) indicates that 52 percent of its builder members already use computer estimating systems. This chapter profiles three of these home builders and shows how estimating software has impacted their work. It also provides a variety of useful ideas on how to successfully incorporate and take advantage of estimating software in your own business.

M/I Schottenstein Homes

Specialty: Production-based home building, ranging from starter homes to semicustom and custom
Computer Estimating Installed: 1980s
Previously Used: Manual estimating methods
Number of Homes Per Year: Approximately 3,500
Number of Employees: Approximately 750

M/I Homes is one of the nation's largest home builders, with 13 home-building divisions serving 11 markets across the country. Much of its work is based on standard home designs that not only include options, but also can be customized according to a homeowner's preferences.

The Ohio-based builder has used its system to create estimate templates for all its standard plans. Each time a new plan is introduced, a new template is created. To produce an estimate for a particular client, the company starts with the base house template, and then adds options and area figures, such as the average quantity of driveway concrete based on the standard setback for homes in the subdivision. (Both area figures and options are stored as assemblies in its estimating database.) Finally, the company does a manual takeoff if a client wants walls moved or other custom work. The entire process takes about 30 minutes for a starter home and up to two hours for very customized homes.

A long-time user of computer estimating, M/I Homes is a believer in the accuracy and time savings that estimating systems can offer home builders. Lately, the company is also seeing a significant reduction in costs as some of its divisions move from using lump sums in their estimates to doing their own takeoffs and comparing their unit costs to the estimates trade contractors submit.

According to Mary Evelyn Hammond, business-systems analyst for the company, doing its own takeoff gives M/I Homes a lot more control over trade-contractor costs. For example, a plumbing contractor may misread a plan's specifications and include the wrong kind of sink. Or maybe a trade contractor has included 10 percent waste for materials when, from experience, M/I Homes knows that the waste will be less. These types of issues can be identified and discussed with trade contractors up front, so M/I Homes knows and agrees to what it is paying for.

Several years ago, the company moved from a minicomputer-based estimating system to one that operates on a PC. The conversion took approximately four weeks per division, including training.

Hammond's implementation advice to other home builders is to make sure that all their trade-contractor records are current before they start building their list of materials for the estimating database. Make sure that payment and bid terms are documented and signed by authorized representatives from both your company and the trade contractor, that the scope of work is defined, and that you have a copy of the agreements on what will be supplied. Preplanning is the key to a smooth implementation.

Schmitt Building Contractors

Specialty: High-end custom homes and light commercial
Computer Estimating Installed: 1992
Previously Used: Manual estimating methods
Number of Homes Per Year: 20
Number of Employees: 60–70 (includes office and field staff)

For Schmitt Building Contractors, computer estimating has helped to double the company's sales volume in five years. This is largely because the company can now produce four times as many estimates as it was able to create by hand.

The North Carolina home builder has set up a sophisticated, fully integrated system that links its estimating software to scheduling, purchasing, and job cost accounting. Recently, Schmitt Building began using digitizer software, which has further reduced its takeoffs from blueprints by an additional 30 percent. It once took the company two weeks to pull together an estimate for a client. It can now create a more accurate estimate in less than four hours.

Schmitt Building has created two different databases to use with its estimating system. One is a unit cost "conceptual" database (based on historical costs) that is used to quickly provide potential clients with an early-on estimate. With this database, the company doesn't have to waste time doing a full takeoff of items before a contract is signed. Once the contract is final, Schmitt Building uses another, more detailed database to take off material quantities for ordering.

Overall, its integrated computer system has helped Schmitt Building to streamline its estimating, purchasing, scheduling, and job cost functions. First, estimate items are coded as to their location in the home, then a schedule is created based on those locations. Next, the company produces purchase orders in sections, according to the delivery schedule. To speed up material delivery, a list of purchase orders is given to superintendents and project managers showing them what they need to order for each phase of the job. Finally, it takes about five minutes to import the estimate into a job cost program to track actual costs against the estimate. Everything is integrated, with information flowing from one software application to another as needed, without double entry of data.

Integrating estimating software with other applications has made jobs more efficient, according to Eric Schmitt, the company's estimator. He indicates that 75 percent of the company's planning is done up front, which has resulted in fewer delays and has ensured on-schedule project completion.

For other builders interested in computer estimating, Schmitt offers the following recommendations:

- Look into the availability of prebuilt databases and how applicable they would be for your company. It took Schmitt six months to build a detailed database from scratch. Prebuilt databases can reduce database development efforts significantly.
- If you have no one who can be your in-house computer expert, consider hiring an outside consultant to build your database. It will better your chances of getting the system up and running quickly.
- Think ahead. Determine what you want from your estimating system five years from now, and take those plans into consideration when you first set up the system, especially when you build your database.
- Remember that computerization is a process. It takes time to get where you want to go and you'll always be learning new ways to improve your efficiency. Be patient.

Jim Murphy & Associates

Specialty: High-end custom and commercial
Computer Estimating Installed: 1995
Previously Used: Electronic speadsheet template
Number of Homes Per Year: Approximately 5
Number of Employees: Approximately 50

Jim Murphy & Associates puts a lot of emphasis on customer relations. That's why the company points to credibility as one of the biggest benefits it has received from computerized estimating. That credibility is established early in the building process when the California-based builder puts together a detailed estimate that is often based only on preliminary drawings. As the drawings progress, the estimate is quickly updated, allowing the company to document all cost-related decisions based on what it is told at the time.

Knowing in detail what will go into a house gives the company more negotiating power when setting the home's final price. It also allows the company's project managers to answer any questions the client might have throughout the project. Clients don't always want to see all the details behind the estimate, says Jay True, vice president of Jim Murphy & Associates, but they're there if the client asks.

Quality reports are also key to customer satisfaction, especially when dealing with today's cost-conscious clients. Computer estimating gives the company the flexibility to provide clients with reports tailored to the way they want to see the information. For example, the company can break down the estimate in different ways, summarize information or provide more detail, and show profit separately or allocated across items in the estimate.

Clients are often impressed with how quickly the company can respond to their information requests, says True. One client, for example, wanted access to specific estimate information to do some financial planning. It took True less than an hour to transfer data from his estimating software into a spreadsheet program that his client could easily use to manipulate the information.

True oversees the estimating process for Jim Murphy & Associates and understands firsthand the role that management plays in the successful implementation of a computer estimating system. He advises other managers to get away from their day-to-day duties so they can focus on the implementation process and the direction it will take. You don't have to learn everything about the system, he says, but deciding on the design of your estimating system is something you can't delegate. There's no substitute, he adds, for the boss understanding what he or she wants to accomplish with computer estimating. True also recommends that supervisors begin using the estimating system on smaller jobs, and manually check their work to make sure everything is working correctly.

Sharing Ideas

Hundreds of home builders nationwide use estimating software. Tapping into their experience can be a valuable component to achieving your own success with an estimating system. Whether you pick up the phone, attend an estimating software conference, or join a discussion on one of the various construction and software web sites, ongoing idea-sharing can turn your estimating system into a powerful competitive tool for your company.

Appendix A

Glossary

Assembly: An estimating software feature that allows you to easily and quickly generate all items and costs associated with a door, wall, or other building component by entering dimensions and sometimes specifications for that component.

Audit Trail: A report that displays items in the sequence they were taken off or in order by the date and time they were last changed.

Bill of Materials (BOM): A list of materials included in an estimate. A BOM is typically organized by type of material (for example, lumber, concrete products, and plumbing).

Bid or Budget: See Estimate.

Chart of Accounts: A list of accounts that serves as the structure for an accounting system.

Computer-Aided Design (CAD): A software program that assists in the drawing and revision of building designs.

Conversion Factor: Factor used to convert from one unit of measure to another.

Cost: The dollar value of each item included in the estimate.

Cost Categories: Broad structure used to group similar cost types (for example, labor, materials, equipment, and trade contractors).

Cost Code: The number or identifier assigned to each cost item. Cost codes grouped together form a chart of accounts.

Cost Item: Each element of an estimate that has any quantity calculated or assigned to it, regardless of whether a cost has been assigned.

Database: A collection of trade contractor quotes, material prices, and other homebuilding information that is organized and stored for easy retrieval and use in constructing the estimate.

Digitize: To enter dimensions and quantities into a computer by using a puck pointer or stylus pen to trace over a home's blueprints.

Digitizer: An electronic tablet used with a stylus pen or puck pointer; the position of this pointer is transmitted to the computer. A digitizer can be used to trace over blueprints, thereby transferring dimensions to the computer estimating system.

Division: A major category of work, such as foundations and exterior framing.

Estimate: A listing of all items and costs required to build a home or perform a unit of work as defined by the user. The estimate also includes factors such as overhead, profit, scheduled cost increases, and contingencies.

Estimate Comparison Report: Any report that compares information, such as unit-cost differences, across a number of estimates.

Field Report: Any report provided to project managers or superintendents for use on the job site.

Graphical User Interface (GUI): A technology that allows you to operate software simply by using a mouse to point and click on symbols and text.

Intelligent Assembly: An assembly with built-in logic. If you change one parameter of a building component, such as wall height, the software automatically makes other necessary adjustments, such as changing the stud spacing.

Markup or Add-On: A percentage, value, or dollar amount used to adjust specific costs, subtotals, and totals. For example, sales tax, bond, insurance, and profit might be tracked as markups or add-ons.

Material List: Listing of all material resources in an estimate with quantity, vendor, and cost.

Multiple Document Interface (MDI): A technology that lets you open and work with more than one document (for example, an estimate) at the same time.

Multitasking: The ability to perform more than one task at the same time (for example, run a report while you enter trade contractor quotes).

Multiuser: Simultaneous access to the same software program by more than one user.

Networkable: Indication of compliance with computer-industry networking standards. A network consists of several computers hooked together to share resources and data.

Object-Based CAD: A CAD system that allows the designer to draw using objects—such as doors, windows, and walls—rather than lines, circles, arcs, and other geometric elements. Unlike a line or circle, a CAD object can contain information. For example, a door object can include the door's cost, which way it swings, its supplier, whether it is hollow or solid-core, and other descriptive information.

On-Line Help: Instructions and usage tips that are available on-screen as you use the program.

Open Database Connectivity (ODBC): A standard interface specification that allows you to move information from one software application to another.

Purchase Order: Documents sent to suppliers to authorize the purchase and delivery of materials for an agreed-upon price.

Plans: A set of detailed blueprints or drawings.

Price: The dollar value for which you purchase material or trade contractor services.

Quote Sheets: A form sent to trade contractors and suppliers that requests a price quote for specified materials and services. The same form can be used by a trade contractor to submit quoted prices.

Sequencing: The ability to sort your estimate in a variety of ways on the computer screen. You can view your estimate by material, contractor, or location (for example, floor 1 and 2).

Spreadsheet: A way of presenting items of work using rows and columns.

Takeoff: Process by which dimensions and other measurements are read off a set of plans and listed in an estimate.

Template Estimate: A prebuilt estimate that can be used to create new estimates.

Template Menu: For digitizers, a printed overlay used to select menu items or symbols.

Trade Contractor Contracts: Documents sent to trade contractors to authorize work at an agreed-upon price.

32 Bit: A technology designed to improve software performance, speed, and use of computer memory.

User-Defined Report: A report in which the user specifies what items to include in the report and what fields to show.

Vendor: Provider of a cost item.

What You See Is What You Get (WYSIWYG): The ability to set up and see a document, such as a report, on the computer screen as it will appear when you print it.

Appendix B

Estimating Software Buyer's Checklist

Choosing a computerized estimating system can be a very confusing task for a building company. There are many packages and hardware configurations, with prices that range from the low hundreds of dollars to several thousands.

The Estimating Software Buyer's Checklist was developed not to give you all the answers in choosing a computerized estimating system but primarily to help you ask the right questions. Use the Estimating Software Buyer's Checklist when evaluating the functionality of the estimating software. Remember, not one system is right for all builders. Each company will have specific needs. Take lots of notes as you go through the process of evaluating different systems.

In addition to considering the functionality and features of the software, consider these additional important factors:

- What type of learning curve can you or your company afford? Plan and account for adequate down-time.
- Is the program written in a manner that you can understand and specific to your industry?
- Are the manuals and tutorials of good quality and written in an easy-to-understand manner?
- Is personal training available? How much does the training cost?
- Where is the training held and how often is it held?
- How thorough is the training? Are advanced classes available?

Investigate the type of technical support you will receive. Here are some questions you might want to consider asking the software vendor:

- Is technical phone support available and to what extent? At what cost?

- During what hours is the support staff available (and in what time zone)?
- How many support people are on staff? Do they understand the construction business?
- How are software upgrades handled and how often are they released?
- What is the cost of an additional support contract?

It is very important to identify the type of work that you do and choose a system based on the tools that you need to accomplish the tasks. Don't get carried away by all the "bells and whistles."

You will also want to consider the expandability of the product as it relates to markets in the future. Examples include links to your job cost accounting system, and CAD or scheduling systems. You may want to investigate whether your system has the ability to link up with other markets.

Choosing a computerized estimating system can be a key factor in the productivity of an organization. Your initial selection should be made with care, not only as regards the hardware and software but also as regards the companies that develop and support the system. To ensure your organization's continual expansion and growth, find a vendor with whom your company can form a strong relationship.

CHECKLIST FORM

Part I: Company Address and Contact Information

Company Name: ______________________________

Street Address: ______________________________

City/State/Zip: ______________________________

Telephone: ____________ Fax: ____________ E-mail: ____________

Web Address: ______________________________

Sales Contact Name: ____________ Sales Phone Number: ____________

Part II: Vendor Profile

Years in Business: ______________________________

Is program directed toward the residential building industry? ____________

____

Who is the target market?

- ☐ Small-volume builders (including custom)
- ☐ Remodelers
- ☐ Land developers
- ☐ Large-volume builders
- ☐ Medium-volume builders
- ☐ Architects
- ☐ Design/drafting services

Other (specify): ______________________________

Product Distribution? ☐ Vendor ☐ Value-added reseller (dealer) ☐ Other ____________

Training: ☐ Corporate site ☐ Regional ☐ Customer site ☐ Telephone ☐ CBT & video Other: ____________

Availability of Tech Support: ____________ Fee Based: ____________

Terms of Warranty: ______________________________

Yearly Program Maintenance Charge: $ ____________

Frequency of Updates: ____________ Date of Last Update: ____________

Additional Services: ☐ User group sponsor ☐ Technical newsletter
☐ Electronic forums ☐ Fax back service

Demo (Disk or CD ROM): ☐ Yes ☐ No

Part III: Product Information

Product Name: ____________________

Version: Number of Current Installations: ____________________

Product Description: ____________________

____

____

____

____

____

____

Retail Price (single user): __________ Network/Multiuser Price: __________

Part IV: Required Operating Environment

Minimum Preferred

Processor: ____________________

Memory (RAM): ____________________

Disk Space for Operation: ____________________

Resolution: ____________________

Operating System: ____________________

____

CD-ROM drive: ☐ Required ☐ Supported

☐ Mouse
☐ Laser Printer
☐ Digitizer

☐ Other: ____________________

____

____

Part V: Estimating Features

Features	Yes	No	Comment
Product designed for residential construction			
Product compatible with Windows			
Backup utilities built into system			
Password protection: Can access to application and/or function be given to individual users?			
Tools to reconstruct data damaged through accidental system shutdown			
Database			
Estimate			
Record locking			
DATABASE			
Includes residential database			
Number of cost items allowed			
User-defined numbering allowed			
Field type allowed			
Structure allowed			
Number of cost categories allowed			
Material			
Labor			
Equipment			
Trade contractors			
Other			
Allows repricing of labor costs for regional differences or requirements			
Maximum number of cost categories per cost item			
Maximum cost item dollar amount allowed			
Allows cost items to convert takeoff units and quantities into order units and quantities (e.g., linear feet into board feet)			

Features	Yes	No	Comment
Can store takeoff information in user-defined categories			
Provides user-defined sort criteria			
Allows alternate cost item coding			
Database item prices held independent of estimate prices			
Cost items can include unique waste factors			
Copies, pastes, and moves items and assemblies			
REPORTS			
Included with the system:			
Summary reports			
Detail reports			
User can create custom reports			
User-defined sort criteria			
Includes a report generator			
Allows printing of a material requirements list:			
Without database prices			
With database prices			
ESTIMATE			
Can update previous estimates using the most current database pricing			
Number of estimates allowed			
Number of sort levels within an estimate			
Number of takeoff items allowed in an estimate			
Allows updating of database cost item prices with new estimate prices			
Allows updating of estimate cost item prices with new database prices			

Features	Yes	No	Comment
TAKEOFF FEATURES			
Allows using previous estimate as template for new estimate			
Single-item takeoff allowed			
Simultaneous, multiple cost item takeoff process allowed			
Name for this process?			
Number of cost items allowed for multiple cost item takeoff process			
Ability to create estimate-specific/one-time items			
Ability to easily move one-time items back to database			
Ability to store takeoff information in user-defined categories			
User-defined sort criteria			
MARKUPS			
Allows for a standard set of system-wide markups			
Allows for markup modification at time of takeoff			
OPTIONAL OR ADD-ON MODULES OR ADVANCED FEATURES			
Multiuser (included with the purchase of additional stations)			
Digitizer interface			
Bid analysis			
Optional databases:			
Model estimating capabilities			

Features	Yes	No	Comment
ABILITY TO INTERFACE WITH OTHER PRODUCTS			
CAD interface			
Purchase Order interface			
Scheduling			
Word processing			
Project management			
Spreadsheets			
Other:			

Appendix C

Product Directory

The software packages referred to in this book were selected because of their focus on the residential construction industry and because they represent a variety of features. Their presence in *Computer Estimating for Home Builders* does not constitute a warranty, expressed or implied, on the part of the publishers, NAHB, editors, or reviewers, nor does the absence of a particular package expressly connote criticism of that package.

With this list we have attempted to present the most up-to-date information on computer estimating software. However, due to the rapid changes in technology, the reader should contact the software vendors directly for the latest product information. Particular decisions the reader might make related to specific features of hardware or software should be based on manufacturer's specifications.

Key to product features:

☐ = Features not included with product
☑ = Features included with product
N/A = Information not provided by vendor

BuildSoft

PO Box 13893
Research Triangle Park, NC 27709
800-999-8322
Fax: 919-941-0339
E-mail: *dbuffaloe@buildsoft.com*
Web Address: *www.buildsoft.com*
Sales Contact Name: Mike Bright
Sales Phone Number: 800-999-8322
Years in Business: 10

Program Directed to Residential Building Industry: Yes

Primary Customers: Small-, medium-, and large-volume builders, remodelers, and land developers

Product Distribution: Vendor

Training: Corporate and customer site, regional, and telephone

Technical Support: 8:30 am–8:30 pm (EST)

Terms of Warranty: 60-day money-back guarantee and 90 days maintenance and support

Yearly Program Maintenance Charge: $345-995

Frequency of Updates: Annual

Last Update: 6/97

Additional Services: —

Demo Disk: Yes

Product Name: BuildSoft

Version: 3.5

Number of Current Installations: 3,000+

Retail Price (Single User): $2,995-5,995

Network/Multiuser Price: $1,250 for two users and $625 each additional user

Processor Requirement (Required/Preferred): 486/Pentium

Memory (RAM) Requirement (Required/Preferred): 16MB/32MB

Disk Space for Operation (Required/Preferred): 50MB/100MB

Monitor/Graphics (Required/Preferred): 640×480/800×600

Operating System (Required/Preferred): Windows 3.11/Windows 95 or Windows NT

CD-ROM Drive Supported: Yes

Other Requirements: Mouse, laser printer, digitizer

Product Description

BuildSoft is an integrated residential-construction-management software system that was designed by builders for builders. BuildSoft gives you total control over every aspect of every job, with complete, seamless integrated automation of such tasks as CPM scheduling, historical/takeoff, estimating, purchase orders/work orders, job costing, and accounting. This easy-to-use system includes a new proposal generator and cash-flow analysis reporting. It also integrates job costs and accounts receivable for accurate time and material billings. BuildSoft integrates with FieldPen at the job site to offer full BuildSoft control in the field.

General Features

- ☑ Compatible with Windows interface
- ☐ Backup utilities built in
- ☑ Individual password protection
- ☑ Data reconstruction tools available
- ☑ Record locking

Database Features

- ☑ Residential database
  Number of cost items allowed: Unlimited
- ☑ User-defined numbering allowed
  Number of cost categories allowed: Unlimited
- ☑ Regional repricing of labor costs allowed
  Maximum cost categories per cost item: Unlimited
  Maximum cost item dollar amount allowed: $999,999,999
- ☑ Can convert takeoff units/quantities into order units/quantities
- ☑ Can store takeoff information in user-defined categories
- ☑ User-defined sort criteria
- ☑ Alternate cost item coding allowed
- ☑ Database item prices held independent of estimate prices
- ☑ Cost items include unique waste factors
- ☑ Copies, pastes, and moves items and assemblies

Report Features

- ☑ Summary reports included
- ☑ Detail reports included
- ☑ User can create custom reports
- ☑ Report generator
  Prints a material requirements list:
  - ☑ Without database prices
  - ☑ With database prices

Estimate Features

- ☑ Updates previous estimate using the most current database pricing
  Number of estimates allowed: Unlimited
  Number of sort levels within an estimate: 4
- ☑ Unlimited number of takeoff items allowed in an estimate
- ☑ Allows updating database cost item prices with new estimate prices

Takeoff Features

- ☑ Allows using previous estimate as template for new estimate
- ☑ Single-item takeoff allowed
- N/A Simultaneous, multiple cost item takeoff process allowed
  Number of cost items allowed for multiple cost item takeoff process: Unlimited
- ☑ Creates estimate-specific/one-time items
- ☑ Easily moves one-time items back to database
- ☑ Stores takeoff information in user-defined categories
- ☑ User-defined sort criteria

Markup Features

- ☑ Standard set of system-wide markups allowed
- ☑ Markup modification allowed at time of takeoff

Optional Modules, Add-On Modules, or Advanced Features

- ☑ Multiuser (included with the purchase of additional stations)
- ☑ Digitizer interface
- ☑ Estimate analysis
  Optional databases: Database is all-inclusive
- ☑ Model estimating capabilities

Ability to Interface with Other Products

- ☑ CAD interface
- N/A Purchase Order interface
- N/A Scheduling
- ☑ Word processing
- N/A Project management
- ☑ Spreadsheets
  Other: —

C.D.C.I. Construction Data Control, Inc.

4000 DeKalb Technology Pkwy, Suite 220
Atlanta, GA 30340
770-457-7725
Fax: 770-457-7686
E-mail: *sales@cdci.com*
Web Address: *www.cdci.com*
Sales Contact Name: Dan Jacobs
Sales Phone Number: 800-285-3929
Years in Business: 19

Program Directed to Residential Building Industry: Yes

Primary Customers: Small-, medium-, and large-volume builders and remodelers

Product Distribution: Vendor, value-added reseller

Training: Corporate and customer site, regional

Technical Support: 10:00 am–5:00 p.m. (EST)

Terms of Warranty: 90-day complete warranty

Yearly Program Maintenance Charge: $350

Frequency of Updates: Annual

Last Update: 3/97

Additional Services: Tech newsletter, fax back service

Demo Disk: Yes

Product Name: Bid Team for Windows

Version: 4.0

Number of Current Installations: —

Retail Price (Single User): $3,995-4,995

Network/Multiuser Price: $5,495 and up

Processor Requirement (Required/Preferred): 486/Pentium

Memory (RAM) Requirement (Required/Preferred): 8MB/32MB

Disk Space for Operation (Required/Preferred): 50KB minimum

Monitor/Graphics (Required/Preferred): —

Operating System (Required/Preferred): Windows 95

CD-ROM Drive Supported: Yes

Other Requirements: Mouse, laser printer, digitizer

Product Description

Bid Team for Windows is a versatile estimating tool that provides full multimedia capacities, enabling you to use sounds and pictures to add clarity and precision to your estimates. This multimedia capability allows you to embed diagrams in the system; to scan in manufacturers' pictures, sketches, and diagrams; and to record sound messages. Bid Team for Windows' spreadsheets are designed to give you increased flexibility. The spreadsheet view allows you to see your estimate in summary form so that during takeoff you can visually confirm that you have all the pieces in your estimate.

General Features

- ☑ Compatible with Windows interface
- ☑ Backup utilities built in
- ☑ Individual password protection
- ☑ Data reconstruction tools available
- ☑ Record locking

Database Features

- ☑ Residential database
  Number of cost items allowed: Unlimited
- ☑ User-defined numbering allowed
  Number of cost categories allowed: Unlimited
- ☑ Regional repricing of labor costs allowed
  Maximum cost categories per cost item: Unlimited
  Maximum cost item dollar amount allowed: $999,999,999
- ☑ Can convert takeoff units/quantities into order units/quantities
- ☑ Can store takeoff information in user-defined categories
- ☑ User-defined sort criteria
- ☑ Alternate cost item coding allowed
- ☑ Database item prices held independent of estimate prices
- ☑ Cost items include unique waste factors
- ☑ Copies, pastes, and moves items and assemblies

Report Features

- ☑ Summary reports included
- ☑ Detail reports included
- ☑ User can create custom reports
- ☑ Report generator
  Prints a material requirements list:
  - ☑ Without database prices
  - ☑ With database prices

Estimate Features

- ☑ Updates previous estimate using the most current database pricing
  Number of estimates allowed: Unlimited
  Number of sort levels within an estimate: 3
- ☑ Unlimited number of takeoff items allowed in an estimate
- ☑ Allows updating database cost item prices with new estimate prices

Takeoff Features

- ☑ Allows using previous estimate as template for new estimate
- ☑ Single-item takeoff allowed
- ☑ Simultaneous, multiple cost item takeoff process allowed
  Number of cost items allowed for multiple cost item takeoff process: Unlimited
- ☑ Creates estimate-specific/one-time items
- ☑ Easily moves one-time items back to database
- ☑ Stores takeoff information in user-defined categories
- ☑ User-defined sort criteria

Markup Features

- ☑ Standard set of system-wide markups allowed
- ☑ Markup modification allowed at time of takeoff

Optional Modules, Add-On Modules, or Advanced Features

- ☑ Multiuser (included with the purchase of additional stations)
- ☑ Digitizer interface
- ☑ Estimate analysis
  Optional databases: Residential, R.S. Means, commercial
- ☑ Model estimating capabilities

Ability to Interface with Other Products

- ☑ CAD interface
- ☑ Purchase Order interface
- ☑ Scheduling
- ☑ Word processing
- ☐ Project management
- ☑ Spreadsheets
  Other: Job Cost, CDCI accounting products

Computerized Micro Solutions CMS

7966 Arjons Drive, Ste. 220
San Diego, CA 92126
619-578-2664
Fax: 619-578-2688
E-mail: *CMS@Proest.com*
Web Address: *www.proest.com*
Sales Contact Name: Jeffrey Gerardi
Sales Phone Number: 800-255-7407
Years in Business: 12

Program Directed to Residential Building Industry: Yes

Primary Customers: Small-, medium-, and large-volume builders and remodelers

Product Distribution: Vendor, value-added reseller

Training: Corporate and customer site, and telephone

Technical Support: Free

Terms of Warranty: 30-day money-back guarantee

Yearly Program Maintenance Charge: —

Frequency of Updates: 18 months

Last Update: 4/1/96

Additional Services: Technical newsletter, electronic forums, fax back service

Demo Disk: Yes

Product Name: ProEst Estimating for Windows 95

Version: 7.0

Number of Current Installations: 5,000

Retail Price (Single User): $695-1,995

Network/Multiuser Price: $350-1,000

Processor Requirement (Required/Preferred): Pentium

Memory (RAM) Requirement (Required/Preferred): 16MB

Disk Space for Operation (Required/Preferred): 20MB

Monitor/Graphics (Required/Preferred): VGA

Operating System (Required/Preferred): Windows 3.1/Windows 95

CD-ROM Drive Supported: Yes

Other Requirements: Mouse, laser printer, digitizer

Product Description

ProEst is a detailed cost estimating system for home builders and remodelers. Create bills of material, labor reports, and customer proposals. Works with digitizers and R.S. Means cost data. Download a free demo at *www.Proest.com* or call 800-255-7407.

General Features

☑ Compatible with Windows interface
☐ Backup utilities built in
☐ Individual password protection
☑ Data reconstruction tools available
☑ Record locking

Database Features

☑ Residential database
Number of cost items allowed: 4,000,000
☐ User-defined numbering allowed
Number of cost categories allowed: 5
☑ Regional repricing of labor costs allowed
Maximum cost categories per cost item: 5
Maximum cost item dollar amount allowed: $999,000,000
☑ Can convert takeoff units/quantities into order units/quantities
☑ Can store takeoff information in user-defined categories
☑ User-defined sort criteria
☑ Alternate cost item coding allowed
☑ Database item prices held independent of estimate prices
☑ Cost items include unique waste factors
☑ Copies, pastes, and moves items and assemblies

Report Features

☑ Summary reports included
☑ Detail reports included
☑ User can create custom reports
☑ Report generator
Prints a material requirements list:
  ☑ Without database prices
  ☑ With database prices

Estimate Features

☑ Updates previous estimate using the most current database pricing
Number of estimates allowed: 9,999
Number of sort levels within an estimate: 2
☑ Unlimited number of takeoff items allowed in an estimate
☑ Allows updating database cost item prices with new estimate prices

Takeoff Features

☑ Allows using previous estimate as template for new estimate
☑ Single-item takeoff allowed
☑ Simultaneous, multiple cost item takeoff process allowed
Number of cost items allowed for multiple cost item takeoff process: Unlimited
☑ Creates estimate-specific/one-time items
☑ Easily moves one-time items back to database
☑ Stores takeoff information in user-defined categories
☑ User-defined sort criteria

Markup Features

☑ Standard set of system-wide markups allowed
☑ Markup modification allowed at time of takeoff

Optional Modules, Add-On Modules, or Advanced Features

☑ Multiuser (included with the purchase of additional stations)
☑ Digitizer interface
☑ Estimate analysis
Optional databases: All R.S. Means databases
☑ Model estimating capabilities

Ability to Interface with Other Products

☑ CAD interface
☑ Purchase Order interface
☑ Scheduling
☑ Word processing
☑ Project management
☑ Spreadsheets
Other: Exports ASCII files in user-defined format

Contractors Software Group

175 Mountain View Drive
Gainesville, GA 30501
770-534-0790
Fax: 770-534-0191
E-mail: *SWCSG@mindspring.com*
Web Address: *http://www.mindspring.com/~swcsg*
Sales Contact Name: Sales Department
Sales Phone Number: 770-534-0790
Years in Business: 5

Program Directed to Residential Building Industry: Yes

Primary Customers: Commercial, small-, medium-, and large-volume, remodelers, land developers and subcontractors

Product Distribution: Vendor and value-added reseller

Training: Corporate and customer site, and telephone

Technical Support: Phone, fax, e-mail

Terms of Warranty: Initial warranty, support, and product maintenance included with purchase

Yearly Program Maintenance Charge: $300

Frequency of Updates: 2–3 times per year

Last Update: 7/97

Additional Services: Technical newsletter, fax back service

Demo Disk: Yes

Product Name: TakeOff Plus

Version: 9707

Number of Current Installations: 500

Retail Price (Single User): $1,295

Network/Multiuser Price: $1,895

Processor Requirement (Required/Preferred): 486/Pentium

Memory (RAM) Requirement (Required/ Preferred): 16MB/32MB

Disk Space for Operation (Required/Preferred): 10MB/20MB

Monitor/Graphics (Required/Preferred): 800×600

Operating System (Required/Preferred): Windows 3.1/Windows 95

CD-ROM Drive Supported: Yes

Other Requirements: Mouse, laser printer, digitizer

Product Description

TakeOff Plus is a complete tool for estimating the entire job from grading to punchout. It enables you to estimate everything, including concrete, framing, siding, brick, stucco, and drywall. TakeOff Plus is designed specifically for Microsoft Windows, so information is just a mouse click away. This ready-to-use package comes with an extensive prebuilt database and an unlimited number of estimates, items, and assemblies. Takeoff checklists help ensure that necessary items are included. It automatically creates purchase orders and subcontracts, and can maintain prices for different suppliers and subcontractors. It also links with popular job cost accounting systems such as the company's own Job Accounting Plus accounting system.

General Features

- ☑ Compatible with Windows interface
- ☑ Backup utilities built in
- ☑ Individual password protection
- N/A Data reconstruction tools available
- ☑ Record locking

Database Features

- ☑ Residential database
  Number of cost items allowed: Unlimited
- ☑ User-defined numbering allowed
  Number of cost categories allowed: Unlimited
- ☑ Regional repricing of labor costs allowed
  Maximum cost categories per cost item: Unlimited
  Maximum cost item dollar amount allowed: Over 999 billion
- ☑ Can convert takeoff units/quantities into order units/quantities
- ☑ Can store takeoff information in user-defined categories
- ☑ User-defined sort criteria
- ☑ Alternate cost item coding allowed
- ☑ Database item prices held independent of estimate prices
- ☑ Cost items include unique waste factors
- ☑ Copies, pastes, and moves items and assemblies

Report Features

- ☑ Summary reports included
- ☑ Detail reports included
- ☑ User can create custom reports
- ☐ Report generator
  Prints a material requirements list:
  - ☑ Without database prices
  - ☑ With database prices

Estimate Features

- ☑ Updates previous estimate using the most current database pricing
  Number of estimates allowed: Unlimited
  Number of sort levels within an estimate: Many
- ☑ Unlimited number of takeoff items allowed in an estimate
- ☑ Allows updating database cost item prices with new estimate prices

Takeoff Features

- ☑ Allows using previous estimate as template for new estimate
- ☑ Single-item takeoff allowed
- ☑ Simultaneous, multiple cost item takeoff process allowed
  Number of cost items allowed for multiple cost item takeoff process: Unlimited
- ☑ Creates estimate-specific/one-time items
- ☑ Easily moves one-time items back to database
- ☑ Stores takeoff information in user-defined categories
- ☑ User-defined sort criteria

Markup Features

- ☑ Standard set of system-wide markups allowed
- ☑ Markup modification allowed at time of takeoff

Optional Modules, Add-On Modules, or Advanced Features

- ☑ Multiuser (included with the purchase of additional stations)
- ☑ Digitizer interface
- ☑ Estimate analysis
  Optional databases: Commercial
- ☑ Model estimating capabilities

Ability to Interface with Other Products

- ☑ CAD interface
- ☑ Purchase Order interface
- ☐ Scheduling
- ☑ Word processing
- ☑ Project management
- ☑ Spreadsheets
  Other: Accounting systems

Enterprise Computer Systems, Inc.

One Independence Pointe
Greenville, SC 29615
800-569-6309
Fax: 864-987-6400
E-mail: —
Web Address: *www.ecs-inc.com*
Sales Contact Name: Trina Rossman
Sales Phone Number: 800-569-6309
Years in Business: 20

Program Directed to Residential Building Industry: Yes

Primary Customers: Small-, medium- and large-volume builders, and land developers

Product Distribution: Vendor

Training: Corporate and customer site, and telephone

Technical Support: Toll-free phone

Terms of Warranty: 90-day warranty

Yearly Program Maintenance Charge: $450 for phone and enhancements

Frequency of Updates: Several per year

Last Update: 8/97

Additional Services: Technical newsletter

Demo Disk: Yes

Product Name: Professional Estimating

Version: 6.22

Number of Current Installations: —

Retail Price (Single User): $1,995

Network/Multiuser Price: $2,295

Processor Requirement (Required/Preferred): —

Memory (RAM) Requirement (Required/Preferred): 16MB /32MB

Disk Space for Operation (Required/Preferred): —

Monitor/Graphics (Required/Preferred): —

Operating System (Required/Preferred): Windows 3.1/Windows 95

CD-ROM Drive Supported: Yes

Other Requirements: Mouse, laser printer, digitizer

Product Description

With BMS For Windows Professional Estimating, you will save time, increase customer satisfaction, and improve accuracy by doing takeoffs and estimates for all phases of construction. It can produce takeoffs for even complex roofs in just a few minutes. Every aspect of the BMS software is fully customizable by the user to offer the flexibility you need when performing a takeoff. And the BMS software interfaces with your accounting program to create job budgets, purchase orders, and cost reports. Another benefit is its easy-to-use Windows features such as pull-down menus, full mouse and stylus control, and drill-down features.

General Features

- ☑ Compatible with Windows interface
- ☑ Backup utilities built in
- ☑ Individual password protection
- ☑ Data reconstruction tools available
- ☑ Record locking

Database Features

- ☑ Residential database
  Number of cost items allowed: Unlimited
- ☑ User-defined numbering allowed
  Number of cost categories allowed: Unlimited
- ☑ Regional repricing of labor costs allowed
  Maximum cost categories per cost item: Unlimited
  Maximum cost item dollar amount allowed: $9,999,999
- ☑ Can convert takeoff units/quantities into order units/quantities
- ☑ Can store takeoff information in user-defined categories
- ☑ User-defined sort criteria
- ☑ Alternate cost item coding allowed
- ☑ Database item prices held independent of estimate prices
- ☑ Cost items include unique waste factors
- ☑ Copies, pastes, and moves items and assemblies

Report Features

- ☑ Summary reports included
- ☑ Detail reports included
- ☑ User can create custom reports
- ☑ Report generator
  Prints a material requirements list:
  - ☑ Without database prices
  - ☑ With database prices

Estimate Features

- ☑ Updates previous estimate using the most current database pricing
  Number of estimates allowed: Unlimited
  Number of sort levels within an estimate: Dependent upon level set-up
- ☑ Unlimited number of takeoff items allowed in an estimate
- ☑ Allows updating database cost item prices with new estimate prices

Takeoff Features

- ☑ Allows using previous estimate as template for new estimate
- ☑ Single-item takeoff allowed
- ☑ Simultaneous, multiple cost item takeoff process allowed
  Number of cost items allowed for multiple cost item takeoff process: User setup–common part maintenance
- ☑ Creates estimate-specific/one-time items
- ☑ Easily moves one-time items back to database
- ☑ Stores takeoff information into user-defined categories
- ☑ User-defined sort criteria

Markup Features

- ☑ Standard set of system-wide markups allowed
- ☑ Markup modification allowed at time of takeoff

Optional Modules, Add-On Modules, or Advanced Features

- ☑ Multiuser (included with the purchase of additional stations)
- ☑ Digitizer interface
- ☑ Estimate analysis
  Optional databases: —
- ☑ Model estimating capabilities

Ability to Interface with Other Products

- ☑ CAD interface
- ☑ Purchase Order interface
- ☑ Scheduling
- N/A Word processing
- N/A Project management
- N/A Spreadsheets
- N/A Other

HomeTech

5161 River Road
Bethesda, MD 20816
800-638-8292
Fax: 301-654-0073
E-mail: —
Web Address: *HomeTechonline.com*
Sales Contact Name: Robert Arbacher
Sales Phone Number: —
Years in Business: 31

Program Directed to Residential Building Industry: Yes

Primary Customers: small-, medium-, and large-volume builders, remodelers, and architects

Product Distribution: Vendor

Training: Telephone and online tutorial

Technical Support: Unlimited (free)

Terms of Warranty: 60-day money-back guarantee

Yearly Program Maintenance Charge: $250

Frequency of Updates: Quarterly

Last Update: 7/97

Additional Services: Fax back service

Demo Disk: No

Product Name: HomeTech Advantage

Version: 1.7

Number of Current Installations: 2,500

Retail Price (Single User): $445

Network/Multiuser Price: $150 per station

Processor Requirement (Required/Preferred): 386/Pentium

Memory (RAM) Requirement (Required/Preferred): 8MB /16MB

Disk Space for Operation (Required/Preferred): 8MB

Monitor/Graphics (Required/Preferred): 640×480

Operating System (Required/Preferred): Windows 3.1

CD-ROM Drive Supported: Yes

Other Requirements: Mouse, laser printer

Product Description

HomeTech Advantage is a Windows-based estimating system. It contains a 3,000-item unit cost database so that estimators can quickly select items to be included in the project and put together an estimate for that project. The system costs are modified quarterly for the area the estimator is working in. The system uses the industry-renowned HomeTech cost database.

General Features

- ☑ Compatible with Windows interface
- ☑ Backup utilities built in
- ☐ Individual password protection
- ☑ Data reconstruction tools available
- ☐ Record locking

Database Features

- ☑ Residential database
  Number of cost items allowed: Unlimited
- ☑ User-defined numbering allowed
  Number of cost categories allowed: Unlimited
- ☑ Regional repricing of labor costs allowed
  Maximum cost categories per cost item: 3
  Maximum cost item dollar amount allowed:
  Can convert takeoff units/quantities into order units/quantities
- ☑ Can store takeoff information in user-defined categories
- ☑ User-defined sort criteria
- ☑ Alternate cost item coding allowed
- ☑ Database item prices held independent of estimate prices
- ☐ Cost items include unique waste factors
- ☑ Copies, pastes, and moves items and assemblies

Report Features

- ☑ Summary reports included
- ☑ Detail reports included
- ☐ User can create custom reports
- ☑ Report generator
  Prints a material requirements list:
  - ☑ Without database prices
  - ☑ With database prices

Estimate Features

- ☑ Updates previous estimate using the most current database pricing
  Number of estimates allowed: Unlimited
  Number of sort levels within an estimate: —
- ☑ Unlimited number of takeoff items allowed in an estimate
- ☑ Allows updating database cost item prices with new estimate prices

Takeoff Features

- ☑ Allows using previous estimate as template for new estimate
- ☑ Single-item takeoff allowed
- N/A Simultaneous, multiple cost item takeoff process allowed
- N/A Number of cost items allowed for multiple cost item takeoff process
- ☑ Creates estimate-specific/one-time items
- ☐ Easily moves one-time items back to database
- ☑ Stores takeoff information into user-defined categories
- ☐ User-defined sort criteria

Markup Features

- ☑ Standard set of system-wide markups allowed
- ☑ Markup modification allowed at time of takeoff

Optional Modules, Add-On Modules, or Advanced Features

- ☑ Multiuser (included with the purchase of additional stations)
- ☐ Digitizer interface
- ☐ Estimate analysis
  Optional databases: —
- N/A Model estimating capabilities

Ability to Interface with Other Products

- ☑ CAD interface
- ☑ Purchase Order interface
- ☑ Scheduling
- ☑ Word processing
- ☑ Project management
- ☑ Spreadsheets
  Other: —

Lantron Technologies, Inc.

429 Catalpa Avenue
North Plainfield, NJ 07063-1815
908-769-8930
Fax: 908-769-1899
E-mail: *info@lantron.com*
Web Address: *www.lantron.com*
Sales Contact Name: Tony Rinaldis
Sales Phone Number: 800-236-6726
Years in Business: 10

Program Directed to Residential Building Industry: Yes

Primary Customers: Small- and medium-volume builders and remodelers

Product Distribution: Vendor

Training: Corporate and customer site, regional

Technical Support: 9–5 p.m. (EST)

Terms of Warranty: —

Yearly Program Maintenance Charge: Free

Frequency of Updates: Every two years

Last Update: 9/97

Additional Services: User Group sponsor, technical newsletter, fax back service

Demo Disk: Yes

Product Name: Quantum Leap Estimator (QLE97)

Version: 3

Number of Current Installations: 3,500

Retail Price (Single User): $695

Network/Multiuser Price: —

Processor Requirement (Required/Preferred): 486DX/Pentium 133

Memory (RAM) Requirement (Required/Preferred): 16MB/24–32MB

Disk Space for Operation (Required/Preferred): 24MB/38.5MB

Monitor/Graphics (Required/Preferred): VGA/SVGA

Operating System (Required/Preferred): Windows 95 or Windows NT

CD-ROM Drive Supported: Yes

Other Requirements: Mouse, laser printer, and digitizer

Product Description

The Quantum Leap Estimator is an estimating system that includes all the tools to keep track of your building or remodeling business. The program includes a Lead Manager to track and analyze leads. QLE97 also includes estimate tracking, takeoff estimating, time material estimating, light project scheduling, contract writer, project specification with drawing and picture, and a full contacts database. The daily activities manager keeps track of all to-do's, punchlists, appointments, meetings, weather log and historical data, and letter generation. QLE97 includes a 12,000-item residential database and a 14,000-item CSI database. Users can even create their own custom database. QLE97 also includes links to Chief Architect, 3D Home Architect, and QuickBook Pro Accounting. Future links will include Fast Track Scheduling, MiniCAD 7, and Softplan. All updates can be downloaded for free at Lantron's web site.

General Features

- ☑ Compatible with Windows interface
- ☐ Backup utilities built in
- ☑ Individual password protection
- ☑ Data reconstruction tools available
- ☐ Record locking

Database Features

- ☑ Residential database
  Number of cost items allowed: Unlimited
- ☑ User-defined numbering allowed
  Number of cost categories allowed: Unlimited
- ☑ Regional repricing of labor costs allowed
  Maximum cost categories per cost item: Unlimited
  Maximum cost item dollar amount allowed: Unlimited
- ☐ Can convert takeoff units/quantities into order units/quantities
- ☑ Can store takeoff information in user-defined categories
- ☑ User-defined sort criteria
- ☑ Alternate cost item coding allowed
- ☑ Database item prices held independent of estimate prices
- ☑ Cost items include unique waste factors
- ☑ Copies, pastes, and moves items and assemblies

Report Features

- ☑ Summary reports included
- ☑ Detail reports included
- ☑ User can create custom reports
- ☑ Report generator
  Prints a material requirements list:
  - ☑ Without database prices
  - ☑ With database prices

Estimate Features

- ☑ Updates previous estimate using the most current database pricing
  Number of estimates allowed: Unlimited
  Number of sort levels within an estimate: 8
- ☑ Unlimited number of takeoff items allowed in an estimate
- ☑ Allows updating database cost item prices with new estimate prices

Takeoff Features

- ☑ Allows using previous estimate as template for new estimate
- ☑ Single-item takeoff allowed
- ☑ Simultaneous, multiple cost item takeoff process allowed
  Number of cost items allowed for multiple cost item takeoff process: Unlimited
- ☑ Creates estimate-specific/one-time items
- ☑ Easily moves one-time items back to database
- ☐ Stores takeoff information into user-defined categories
- ☑ User-defined sort criteria

Markup Features

- ☑ Standard set of system-wide markups allowed
- ☑ Markup modification allowed at time of takeoff

Optional Modules, Add-On Modules, or Advanced Features

- ☑ Multiuser (included with the purchase of additional stations)
- ☐ Digitizer interface
- ☑ Estimate analysis
  Optional databases: CSI, Residential Database, and a blank create-your-own.
- ☑ Model estimating capabilities

Ability to Interface with Other Products

- ☑ CAD interface
- ☑ Purchase Order interface
- ☑ Scheduling
- ☑ Word processing
- ☑ Project management
- ☑ Spreadsheets
  Other: Links to MiniCAD and Softplan

Timberline Software Corporation

9600 SW Nimbus Avenue
Beaverton, OR 97008
503-626-6775
Fax: 503-526-8010
E-mail: *precision.info@timberline.com*
Web Address: *www.timberline.com*
Sales Contact Name: Jan Petersen
Sales Phone Number: 800-628-6583
Years in Business: 26

Program Directed to Residential Building Industry: Yes

Primary Customers: Medium- and large-volume builders, and architects

Product Distribution: Value-added reseller

Training: Corporate site, regional

Technical Support: Toll-free, US fee-based phone; combined support/maintenance plan and online technical support

Terms of Warranty: 90 days free support and maintenance

Yearly Program Maintenance Charge: Based on flat fee and percentage of software owned

Frequency of Updates: Averages one a year

Last Update: Fall 1997

Additional Services: User group sponsor, technical newsletter, electronics forums, fax back service

Demo Disk: Yes

Product Name: The Precision Collection

Version: 2.0

Number of Current Installations: 18,000 workstations worldwide

Retail Price (Single User): —

Network/Multiuser Price: —

Processor Requirement (Required/Preferred): 486/Pentium

Memory (RAM) Requirement (Required/Preferred): 16MB/24MB

Disk Space for Operation (Required/Preferred): 15MB/50MB

Monitor/Graphics (Required/Preferred): 640×480/800×600

Operating System (Required/Preferred): Windows 95 or Windows NT

CD-ROM Drive Supported: Yes

Other Requirements: Mouse; digitizer optional

Product Description

The Precision Collection is a complete line of estimating software that you can use to produce everything from a preliminary estimate to a complete bill of materials. Designed specifically for Microsoft Windows 95 and NT, Precision's core software incorporates the latest ease-of-use features to let you point-and-click and drag-and-drop your way through estimate and budget creation. A variety of additional software is available to further improve your efficiency, including links to popular CAD and scheduling software.

General Features

☑ Compatible with Windows interface
☑ Backup utilities built in
☑ Individual password protection
☑ Data reconstruction tools available
☑ Record locking

Database Features

☑ Residential database
Number of cost items allowed: Unlimited
☑ User-defined numbering allowed
Number of cost categories allowed: 5
☑ Regional repricing of labor costs allowed
Maximum cost categories per cost item: 3
Maximum cost item dollar-amount allowed: $99,999,999.99
☑ Can convert takeoff units/quantities into order units/quantities
☑ Can store takeoff information in user-defined categories
☑ User-defined sort criteria
☑ Alternate cost item coding allowed
☑ Database item prices held independent of estimate prices
☑ Cost items include unique waste factors
☑ Copies, pastes, and moves items and assemblies

Report Features

☑ Summary reports included
☑ Detail reports included
☐ User can create custom reports
☐ Report generator
Prints a material requirements list:
☑ Without database prices
☑ With database prices

Estimate Features

☑ Updates previous estimate using the most current database pricing
Number of estimates allowed: Unlimited
Number of sort levels within an estimate: 3
☑ Unlimited number of takeoff items allowed in an estimate
☑ Allows updating database cost item prices with new estimate prices

Takeoff Features

☑ Allows using previous estimate as template for new estimate
☑ Single-item takeoff allowed
☑ Simultaneous, multiple cost item takeoff process allowed
Number of cost items allowed for multiple cost item takeoff process: Unlimited
☑ Creates estimate-specific/one-time items
☑ Easily moves one-time items back to database
☑ Stores takeoff information in user-defined categories
☑ User-defined sort criteria

Markup Features

☑ Standard set of system-wide markups allowed
☑ Markup modification allowed at time of takeoff

Optional Modules, Add-On Modules, or Advanced Features

☑ Multiuser (included with the purchase of additional stations)
☑ Digitizer interface
☑ Estimate analysis
Optional databases: Home Builder, Commercial G.C., HVAC, Plumbing, Concrete-Masonry, Electrical, Industrial and Plant
☐ Model estimating capabilities

Ability to Interface With Other Products

☑ CAD interface
☑ Purchase Order interface
☑ Scheduling
☑ Word processing
☑ Project management
☑ Spreadsheets
Other: ODBC compatibility, links to reporting and analysis software, and interfaces to R.S. Means databases